BROODSTOCK MANAGEMENT AND FISH SEED PRODUCTION

The book entitled "Broodstock Management and Fish Seed Production" provides information relating to commercially cultivable fresh water fishes, broodstock management, fish seed production technology, fish seed quality management including induced breeding with neat illustration. Increasing the aquaculture production can be achieved through the supply of quality fish seed. This book will be immensely helpful to farmers, hatchery managers, entrepreneurs and fisheries graduates pursuing research in the area of freshwater broodstock management and sustainable development of freshwater aquaculture in the country.

Dr. K. Karal Marx is basically a Fisheries Biotechnologist mainly involved in teaching "Fish genetics and breeding" and "Fisheries Biotechnology" subjects to UG, PG and Ph.D. students. He has worked as Professor & Head, Department of Fisheries Biotechnology at Fisheries College and Research Institute, Thoothukudi.

Dr. J.K. Sundaray, presently working as Head, Division of Fish Genetics and Biotechnology at ICAR- Central Institute of Freshwater Aquaculture, Kausalayaganga, Bhubaneswar. At present he is associated with the project in relation to selective breeding of Indian Freshwater Fishes, Marker Assisted Selection, Transgenic and Molecular Endocrinology.

A. Rathipriya, graduate in Fisheries Science and post graduate with the specialization in 'Fish Biotechnology' at Fisheries College and Research Institute - Thoothukudi, Tamil Nadu Dr J. Jayalalithaa Fisheries University, Nagapattinam. She is presently doing Ph.D in Fish Biotechnology at TNJFU. She has served as Senior Research Fellow in NADP, and TANII Projects in TNJFU.

M. Muthu Abishag is a post graduate research scholar in Tamil Nadu Dr. J. Jayalalithaa Fisheries University. With a passion for aquaculture research, she now works in the gametology and cryopreservation of fish gametes. She is the recipient of "Dr. K.C. Naik Award" for securing first rank in the all India level ranking for Best Fisheries Graduate of India during 2017.

BROODSTOCK MANAGEMENT AND FISH SEED PRODUCTION

K. Karal Marx
Dean,
Institute of Fisheries Post Graduate Studies
Tamil Nadu Dr. J. Jayalalithaa Fisheries University
OMR campus , Chennai – 603103

A. Rathipriya
Ph.D (Fish Biotechnology)
Institute of Fisheries Post graduate studies
Tamil Nadu Dr. J. Jayalalithaa Fisheries University
OMR campus , Chennai - 603103

J.K. Sundaray
Head,
Division of Fish Genetics and Biotechnology
ICAR-Central Institute of Freshwater Aquaculture
Kausalyaganga, Bhubaneswar-751002, ODISHA

M. Muthu Abishag
M.F.Sc., (Aquaculture)
Fisheries College and Research Institute,
Tamil Nadu Dr. J. Jayalalithaa Fisheries University
Thoothukudi-628008

CRC Press
Taylor & Francis Group
Boca Raton London New York

CRC Press is an imprint of the
Taylor & Francis Group, an **informa** business

NARENDRA PUBLISHING HOUSE
DELHI (INDIA)

First published 2021
by CRC Press
2 Park Square, Milton Park, Abingdon, Oxon, OX14 4RN
and by CRC Press
6000 Broken Sound Parkway NW, Suite 300, Boca Raton, FL 33487-2742

CRC Press is an imprint of Informa UK Limited

Print edition not for sale in South Asia (India, Sri Lanka, Nepal, Bangladesh, Pakistan or Bhutan).

British Library Cataloguing-in-Publication Data
A catalogue record for this book is available from the British Library
Library of Congress Cataloging-in-Publication Data
A catalog record has been requested

ISBN: 978-0-367-62971-7 (hbk)
ISBN: 978-1-003-11168-9 (ebk)

**NARENDRA PUBLISHING
HOUSE
DELHI (INDIA)**

CONTENTS

Foreword   *vii*

Preface   *ix*

1. Candidate Fish Species for Aquaculture   1

2. Setting up a Fish Hatchery   15

3. Transportation of Fish Brooders and Fish Seeds   25

4. Maintenance and Handling of Brooders   30

5. Induced Breeding of Carps   36

6. Significance of Quality Seeds in Sustainable Fish Production 56

7. Do's and Don'ts in the Production of Fish Seeds   64

8. CIFABROOD™ Feed as a Healthy Feed for Brood Fishes 68

9. Induced Breeding and *in vitro* Fertilization in   71
    Common Carp

10. Farming of Loaches with Carps   79

11. Induced Breeding and Seed Production in Catfishes   83

12. Broodstock Development, Seed Production and Hatchery   96
    Management of Tilapia

13. Water Quality Management in Aquaculture   119

14. Fish Diseases Treatment and its Preventive Measures   135

    References   146

TAMIL NADU Dr. J. JAYALALITHAA FISHERIES UNIVERSITY

Estd. G.O(Ms) No: 118 AH,D&F, (Jul.12)
Accredited under NAEAB, ICAR (Nov.17)
Listed under Section 12B of UGC Act, 1956 – 09.11.2013 (Jan.18)

Dr. S. Felix, M.F.Sc., Ph.D.,
Vice-Chancellor

Vettar River View Campus
Nagapattinam – 611 002
Tamil Nadu, India

Date: 02.04.2019

FOREWORD

Fish has a major role in the income generation of farmers and being the simple food for the poor. In the global scenario, India stands second in fish production next to china. This is mainly due to the successful implementation of 'induced fish breeding' and "composite carp culture' technology by the Central and state governments across the country. There are over 1500 hatcheries established in the country, producing over 32 billion carp fry. Over the years the hatcheries are functioning as isolated units without exchange of brooders that resulted in the deterioration of the quality seed. Unfortunately, the lack of genetic improvement has resulted in the production of poor quality seeds in many fish hatcheries in aquaculture, it is essential to produce good quality fish seeds in large quantities, especially of the carps, since carp culture forms the mainstay in Indian aquaculture and contribute as much as 87% of the total aquaculture production. Moreover, nearly 5-10% of the production cost is spent on the purchase of fish seeds. But lack of production of good quality seeds for the culture practices has resulted in reduced growth and production, seriously affecting the income and the economics of the fish farmers. Considering the present scenario and keeping the importance of quality fish seed the authors have written a book on, **"Broodstock Management and Quality Fish Seed Production in Freshwater Fishes".** I hope this book will be immensely useful for the farmers, hatchery   owners and also students. I congratulate Prof. Dr. K. Karal Marx and his co-authors for bringing out this book at the right time.

(S. FELIX)

Phone: +91 4365 256444, Fax: +91 4365 256443   Email: vc@tnjfu.ac.in / felix@tnjfu.ac.in,
Website: www.tnjfu.ac.in

PREFACE

Aquaculture forms one of the main food production activity through out the world. At present the total capture fisheries production is declining and aquaculture production is increasing. Next to China, India ranks second in inland aquaculture production. Although different technologies are available for increasing the fish production further increase in production is possible through quality fish seed production. The fish hatchery managers mainly focus on the quantity of seeds produced and not on the quality of the seeds. Hence, it is felt adequate knowledge on the brood stock management, maintenance of effective breeding population size to avoid inbreeding, cross breeding and obtaining genetic gain are lacking among the hatchery operators. In order to create awareness among the farmers and hatchery owners the authors have written this book. This book includes chapters covering almost all aspects of quality fish seed production in freshwater fishes. We hope that the reader who is interested in farming of freshwater fishes will find this book very useful practically for the production of quality fish seed. We sincerely thank M/s Chola Fish Seed Hatchery, Vaduvoor, and M/s Arvinth Fish Farm, Thirumeni Aeri, Mannargudi Taluk, Thiruvarur District, Tamil Nadu for providing valuable information about their experience in fish seed production. Finally, we express our gratitude to our Vice-Chancellor, Tamil Nadu Dr J. Jayalalithaa Fisheries University, for the enormous, continuous support and encouragement given for the preparation of this book.

K. Karal Marx
J.K. Sundaray
A. Rathipriya
M. Muthu Abishag

1

---•◆•---

CANDIDATE FISH SPECIES FOR AQUACULTURE

There are about 30,700 identified fish species in the world. India has a diversity of 2246 fish species. Of that, 37 species are potential for culture. Since the successful fish farming depends on the selection of suitable species, knowledge on the candidate species for culture is necessary.

1. Catla

This species was first introduced in 1921 in Tamil Nadu and popularly known as "Thoppa "in Tamil. Scientific name is *Catla catla* (Fig 1.1). It is the fastest growing fish among the Indian Major carps. It is very well distinguished by its larger head, superior mouth and its broader body shape in lateral compression. The body is greyish white in colour with the reddish operculum area and the dorsal and caudal fins show a blackish tinge. It feeds on zooplankton of the pond surface using large gill rakers. With the supplementary feed given, it can reach up to 1-1.5 kg in the first year. Males matures in the second year and females matures at the end of third year and breed naturally in the rivers and canals during the south west monsoon months of June, July and August. It grows to a maximum of 45 kg in weight and 5 feet in length in wild environment.

Fig. 1.1. Catla

2. Rohu

Scientific name is *Labeo rohita* (Fig 1.2). Of all the carps, this is considered as the tastiest fish. It has a small and pointed head, terminal small mouth with fringed lower lip. It is identified with its dull reddish

Fig. 1.2. Rohu

scales. It is a column feeder on phytoplankton, plant debris on decaying aquatic plants. However, it feeds on zooplankton in its younger stage. It reaches about 1 kg in the first year and attains sexual maturity in the second year and breeds similar to that of catla. It can reach a maximum of 3 feet in length and 30 kg in weight.

3. Mrigal

It is the third biggest group in the Indian Major Carps. The scientific name of mrigal is *Cirrhinus mrigala* (Fig 1.3). It has a linear body, a small head with blunt snout and a sub terminal mouth with thin non-fringed lips. The mouth appears to be superior since the lower lip is shorter than the upper. It is a bottom feeder on decaying organic and vegetable debris. However, its young feed on zooplankton. It grows to about 700 gms to 1 kg in the first year. The time of maturity and breeding are similar to that of rohu. It can reach a maximum of 30 kg and grows to about 3-3.5 feet long.

Fig.1.3. Mrigal

4. Silver carp

It is a Chinese carp introduced in 1959 by importing from Hong Kong and Japan. The scientific name of silver carp is *Hypophthalmichthys molitrix* (Fig 1.4). An upturned mouth and laterally compressed body with small silvery scales are its identifying features. It is a surface feeder and mainly feeds on phytoplankton by straining using their gill rakers. Because of its feeding habits and the complete digestion and utilization of food, it grows faster than catla and can reach up to 1.5 to 2.0 kg in the first year. Under Indian climatic conditions sexual maturity takes place in the second year. Breeding occurs during the months of June to August in natural waters but not in the lakes and ponds.

Fig. 1.4. Silver carp

5. Grass carp

Grass carp is also a Chinese carp that was imported from Japan and introduced in India in 1959. The scientific name of grass carp is *Ctenopharyngodon idella* (Fig 1.5). It has an elongated body, broad head with rounded snout, slightly longer upper jaw and moderately sized scales with light – greenish tinge. It has acquired this name since it preferentially feeds on aquatic plants except *Eicchornia*, aquatic weeds and terrestrial grass (Napier grass). It manures the fish ponds as fifty percent of the daily food consumed, which constitutes organic manure. Normally it reaches 2-3 kg in the first year. The maturity and breeding habits are similar to that of silver carp. This fish is known as 'sanitary fish' since it controls aquatic weed growth in fish ponds.

Fig. 1.5. Grass carp

6. Common carp

It is a Chinese carp introduced in India in 1939 and 1957 from Srilanka and Thailand. The scientific name of common carp is *Cyprinus carpio* (Fig 1.6). It is a freshwater fish growing well in lakes and ponds. It grows well in the tropical and sub – tropical countries and acclimatizes easily to its habitat. Among the various species of culturable fishes, common carp is the most widely and largely cultured fish in the world. In China, common carp is the first fish to be acclimatized for culture in fish farms. Later it spread to other parts of the world and widely introduced and launched in Asian and European countries. Common carp is one of the important species included in the polyculture and with composite fish culture system along with Indian Major Carps.

Fig. 1.6. Common carp

The common carp reach about 1 kg in the first and attain maturity in the same year in tropical countries like India. The eggs are small and adhesive in nature and it breeds throughout the year but mostly bred during the month of July to August, February and March. They can be bred by natural or artificial breeding method.

There are three varieties of common carp,

i.   Mirror carp

ii. Leather or nude carp

iii. Scaled carp

(i) **Mirror carp:** The scales of mirror carp are large, shiny like a mirror but are irregular and scattered.

(ii) **Leather / Nude carp:** It is distinguished by the leathery appearance due to the absence of scales. However, a few scales may be present in some fishes.

(iii) **Scaled carp:** It is characterized by its small scales which will be regularly arranged in rows covering the entire body surface.While scale carp and mirror carp thrive in plains, leather carp is mostly confined to cold upland waters. It is an omnivorous bottom feeder on larvae of insects, worms, mollusc and detritus and leaves of submerged plants. It can reach up to 1-1.5 kg in the first year. But it is known for damaging the dikes of the pond by digging it.

7. Catfish

About 45 species of catfishes are widely spread from Africa to Philippines. Among them, the species *Clarias magur* (Fig 1.7) is known as the 'Asian catfish' or Magur in North India. It is commonly found in the fresh and brackish water areas of India, Myanmar, Sri Lanka, Pakistan, Bangladesh, Malaysia, Thailand, Indonesia, Taiwan and China. It inhabits the natural water bodies like rivers, canals and lakes.

Fig. 1.7 Catfish (Magur)

It is identified by its elongated cylindrical body with a depressed head and very small eyes protruded above the head. It has four pair of barbells around its mouth, hence commonly called as 'catfish'. The dorsal fin will be long and spineless extending upto the caudal fin. The pectoral fins will be well developed with spines used as defense when attacked producing severe pain. It is capable of directly breathing atmospheric air due to the presence of accessory respiratory organs and hence can live in water bodies with low atmospheric oxygen. It is also known as 'walking catfish' and can move from one pond to another during rainy season.

It reaches about 200 – 250 gms in 6-8 months. Males and females can be distinguished by secondary sexual characters after they attain maturity. The females appear dark grey while the males will be brownish in appearance. The males would be smaller than females and occur in the ratio of 1:2. The abdomen of a gravid female would be round, bulging with a reddish colored vent having round and button shaped genital papilla and the males have elongated and pointed papilla (Fig. 11.2 and 11.3). They attain maturity at the end of first year in tropical countries and one and a half or the second year in subtropical and temperate countries. Hence the climatic conditions play a major role in their sexual maturity. The fecundity varies between 15,000 – 20,000.

8. *Pangasius* sp

Panagasianodon aquaculture is the fastest growing food production sector in the world. *Pangasianodon hypothalamus* a native of Mekong River in Vietnam has been introduced to many Asian countries such as Singapore, Philippines, Taiwan, Malaysia, China, Myanmar, Bangladesh,

Fig. 1.8 Pangasius

Nepal including India. The body will be elongated, laterally compressed without scales. There will be 4 pairs of maxillary barbells and preferentially a carnivore. It reaches about 3-4 kg in second year. Usually, the females attain maturity in the third year and the males in the second year. It breeds during June to September and sometimes till November naturally in rivers. The eggs are adhesive is nature. *P. hypophthalmus* is being cultured in many states particularly Andhra Pradesh and West Bengal form the hub of seed production of this species. Over 500 crores of seed is being produced every year. The farming permission was issued in 2011 by Government of India.

9. Tilapia

There are more than 100 species of Tilapia in the world and among them two species; Nile Tilapia and Red Tilapia have gained aquaculture importance. The countries like, Israel, China, Egypt, Indonesia, Philippines, Thailand and Vietnam culture Tilapia in large quantities. The global production of Tilapia is about 6.5 million metric tonnes, one – third of which is contributed by Asian countries. Tilapia is farmed in more than 140 countries across the globe. China being world's largest producer, consumer and exporter.

The yield depends on the extent of culture and the production will be 2 tonnes per hectare by extensive culture method, 8 tonnes/ha by semi – intensive culture and 10 to 20 tonnes by intensive culture with the provision of supplementary feed and aeration. With the supply of supplementary feed, aeration and water exchange, the production could be increased to 20 to 200 tonnes/ha in ultra-intensive culture. In India, it can be cultured with the proper permission from the government after registration and there is a growing interest among the farmers in Tilapia culture.

(i) **Nile Tilapia:** It is characterized by vertical black stripes in the body and reddish tinge in the caudal fin. The contribution of Nile Tilapia (Fig 1.9) to the global production of Tilapia is about 70%. Moreover, there are genetically improved varieties of Nile tilapia like the Israeli strain, Chithralatha strain and GIFT by World fish (Genetically Improved Farmed Tilapia). It is also known as St.Peters fish.

Fig. 1.9. Nile tilapia

(ii) **Red Tilapia:** It also grows faster like the Nile Tilapia and is most suitable for culture in estuarine ponds and sea cages. It is produced by the hybridization of *Oreochromis mossambicus* (Fig 1.10) and *O. niloticus*. It is known as Taiwan red Tilapia. The Philippine red tilapia is produced by hybridizing *Oreochromis mossambicus hornorum* x *Oreochromis niloticus*. There is a great demand for live or frozen red tilapia fish in South – East Asian countries. It is sold in the name of "Emperor Fish" or "Pearl Fish" in the restaurants of Philippines. It has got a good export market and is sold as fillets or other value added products in Japan and America.

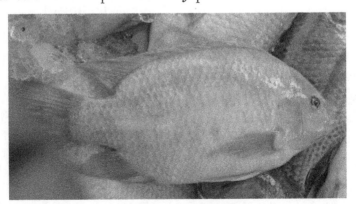

Fig. 1.10. Red Tilapia

10. Snakeheads or Murrels

Snakehead or murrels (Fig 1.11) have long been regarded as a valuable food fish by the Asian people. In India, Thailand, Malaysia and Vietnam it is one of the most common staple food fishes. In the past most of the supply of snakehead came from capture fisheries. In recent years, however, declines in the harvests of wild snakehead, due to overfishing and the destruction of spawning areas with pollution from industrialization, along with the suitability of the fish for culture by virtue of its air breathing characteristic and hardiness, have created keen interest in farming the species. There are several species of snakehead belonging to the genus *Channa*, but only one species *Channa striatus* is cultivated. Commonly called as striped murrel. Many farmers show interest towards *C. striatus* for various reasons. Murrels are highly delicious, air breathing and can be stocked in high stocking densities. Can be sold in live condition and fetch higher price among the fresh water fishes.

Fig. 1.11. Stripped murrel

The seed availability is the major constraint for murrel farming. At present the farms obtain their fry for stocking from specialist fry collectors. These collectors, using their knowledge of the natural spawning grounds, collect wild fry using simple netting procedures. Most of the farmers stock their ponds using fry of more than 10 days of age and about 1 – 2 cm in size. Very few farmers stock their ponds with fingerling size as such, since there are no specialist fingerling producers. There is some artificial spawning of *C. striatus* carried out. The brood stocks were provided with as near natural conditions as possible. They are usually placed in ponds with muddy bottoms and in very shallow waters of between 30 – 100 cm deep. Aquatic vegetation is introduced, (usually water hyacinth) and fish placed in such ponds normally spawn in one to two months time. The raising period ranges from 7 to 10 months depending on the types of system, the stocking densities and the feed used. The fish at harvesting time weigh between 700 – 1,000 gm. About 15-25 tons/ha can be produced in Thailand. Because of the lack of domestication and controlled breeding seeds are not available to farmers in India. But there is lot of scope for murrel farming in India.

11. Pabda- *Ompok bimaculatus*

It has two pairs of barbells; the maxillary pair is longer in length reaching the anal fin while the mandibular pair is short, sometimes rudimentary. The rayed dorsal fin is shorter while the pectoral fin is long and surpasses the pelvic fin reaching to the anal fin. The caudal fin is deeply forked with pointed lobes (the upper lobe conspicuously longer than the lower). The skin is smooth and silvery with a purplish tinge. The dorsal portion has faint shade of a darker green with a yellow tinge. The fins are pale gold in colour. There is a small triangular black spot just above the lateral line and the caudal peduncle. The maximum reported size is 45.7 cm. *O. bimaculatus* (Fig 1.12) attains maturity during first year. Males mature by late April while females attain full growth from late

May to the end of July. The breeding season of the fish extends from early June to late July. However, in places where the monsoon comes early such as in North eastern part of India, breeding periods range from April to July.

Fig. 1.12. Matured pabda

12. *Mystus gulio*

Bagrid catfishes of genus *Mystus* are widespread, but native to Asia. There are 45 species known belonging to the genus *Mystus*. In many Indian states, they fetch high price in markets compared to other small indigenous fish species (SIFs). Among 19 *Mystus* species reported in India, *M. gulio* (Fig 1.13) is considered as one of the candidate species for commercial farming due to its faster growth and size. The maximum size of the species reported is 46 cm. *M. gulio* matures at the age of one year. Breeding season extends from early June to late August. Absolute fecundity of *M. gulio* ranges from 4,000 to 1,70,000 with an average of 32,000. The average number of ova per g ovary ranges from 2,600 – 2,800. The avg. number of ova per g body weight is 755. The fully developed ova are light orange to dull brown in colour, and measures 0.60-0.65 mm in diameter. The average water hardened egg diameter is 0.95-1.25 mm. The length of newly hatched out larvae ranges from 3.20 – 3.8 mm.

Fig. 1.13 *Mystus gulio* adult female

2

SETTING UP A FISH HATCHERY

Selection of suitable site and setting up of infrastructure facilities is a prerequisite for the establishment of a profitable fish hatchery. The number and the size of the ponds should be designed depending on the species used for seed production.

Site selection for a fish seed hatchery

A viable fish culture practice primarily depends on the selection of a suitable site which in turn depends upon water retention, quality of the soil and availability of adequate water supply for hatchery operation. Gravelly and sandy soils have poor water retaining capacity and high rates of seepage and are not suitable for fish culture. On the other hand, the areas with clayey or clay loam soil retain water to a greater extend and are most preferred for construction of fish ponds. It is always better to determine the nature of soil down to a depth of about 1m in the proposed site for pond construction. In order to get adequate water for the fish ponds, the areas for the construction of these ponds can be near stream, canal, ponds or reservoir. Wells and bore wells can also be used as a source of water, if sufficient quantity of water can be obtained from them. While selecting the site for the construction fish hatchery water quality analysis should be done in the proposed site. The pH of the water should be of the range 7-9 and the organic carbon content should be 1.5-2.5%. Site with uniform natural slope towards one side are suitable for pond construction. Moreover, the proposed site should have the proper infrastructure for all logistic and basic facilities for the farm personnel.

1. Designing of the ponds

The ponds can be constructed as per our convenience (Fig 2.1). Usually the production ponds should be larger for carps covering about 200 to 1000 m^2. It is always economical if the bundh is formed out of the earth removed by digging the pond. These bundhs help in holding water inside the ponds. The bundhs are designed such that the top portion is flat with slopes on both the sides so that the bundh does not collapse due to higher water pressure. The ratio of inner (wet) and the outer (dry) slopes should be 1.5:1 and 1:1 respectively (Fig 2.2). The height of the bundh should be 0.25m greater than the depth of water depending on the type of soil. The water depth can be maintained upto 1m. But, such greater depth is not necessary for catfish, pabda, mystus and murrel culture, since they are air breathing.

Fig. 2.1. Fish culture pond

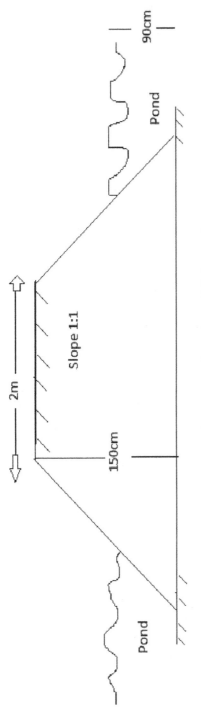

Fig. 2.2. Cross- section view of fish pond

The inlet canals take water to the ponds from the water source. Usually the primary inlet canal would be constructed at the centre of the farm and water will be supplied to the ponds on both of its sides using secondary inlet canals. The drainage canals can be constructed around the periphery of the farm. A sluice or a monk can be used to regulate the water flow at the outlet. They also help in preventing the escape of cultured fishes from the pond.

2. Preparation of the ponds

The pond bottom should be tilled (harrow) during the drying period to increase the oxygen content of the soil, especially if it has a heavy texture (clays and clay loams). A disc harrow is the best equipment to use and tilling should take place while the soil is still wet but is dry enough to support the weight of the tractor (Fig 2.3). Where there has been a severe disease problem in the previous crop, should spread 1,000 kg/ha of agricultural limestone ($CaCO_3$) or 1,500 kg/ha of hydrated lime [sometimes called slaked lime – $Ca(OH)_2$]. It is better if agricultural limestone is used. The use of slaked lime, or quick lime (CaO) may increase the subsequent pH of the water above tolerance

Fig. 2.3. Pond preparation

limits if fishes are stocked soon after the ponds are filled. After adding agricultural limestone we should sun-dry the ponds for at least two weeks so that toxic gases such as hydrogen sulphide and methane are avoided. Some freshwater prawn farms make a standard application of 1000 kg/ha of agricultural limestone every time a pond is drained. Chlorination can also be used for disinfection but it is not recommended because it is a much more expensive treatment. If the ponds are undrainable mahua oil cake (MOC) can be applied at the rate of 250 kg/ha to remove the unwanted fishes. They will completely eradicate weed fishes in one or two days and the MOC will serve as natural manure thereafter. Usually the toxicity lasts for about 15 days only. After harvesting the pond, the water should be drained completely before stocking the next crop. The ponds should be sun-dried and ploughed so that the productivity can be increased

3. Liming

Measuring soil pH

Take 10-12 samples of the upper 5 cm layer of the soil, before any soil treatment has been applied, dry them in an oven at 60°C, and pulverize them to pass a 0.085 mm screen. Bulk the samples together and mix 15 g of the pulverized soil with 15 ml of distilled water. Stir occasionally for 20 minutes and measure the pH, preferably with a glass electrode. The hand-held pH-soil moisture testers used by some farmers are not accurate enough.

The pH suitable for fish culture is 7.5-8.5. The pH should be identified and if it is acidic, then liming can be done to correct it. Depending on the pH of the pond, liming can be done as follows (Table 2.1). Liming is an important step in the preparation of fish culture ponds. If the pH of the pond is in the range 5.1 to 6.9, a minimum of 600 kg to maximum of 1000 kg lime per hectare can be applied to correct the pH.

Table 2.1. Quantity of lime to be applied for a hectare

Sl.No.	pH of the soil	Quantity of lime (Kg)
1	4.6 – 5.0	2000
2	5.1 – 6.5	1000
3	7.6 – 8.5	200

4. Manuring

The ponds can be filled with water to the desired depth. Ten days after liming, the ponds must be fertilized for the propagation of natural food of fishes like the plankton. The manure from cattle, sheep, pig and chicken can be used as organic manure. Cow dung is most preferable among the organic manures. It can be applied at the rate of 10 tonnes/ha/year depending on the water quality. One-sixth of this amount must be diluted in water and applied as a basal dose 10 – 15 days before stocking. If mahua oil cake had been used earlier at 250 kg/ha, there is no need for this basal dose. The remaining amount is applied in equal installments every month. This time they must not be diluted in water and must be dumped directly into the pond at different locations every month.

5. Use of synthetic fertilizers

Besides the organic manures, inorganic fertilizers can be applied every year in a one hectare pond at the rate of 200 kg Urea, 250 kg Super Phosphate and 40 kg Potassium. As done for organic manures, one-sixth of the amount of inorganic fertilizers must be applied as a basal dose before stocking. The remaining amount can be applied every month in equal amount. Both the organic and inorganic fertilizers must not be applied at the same day but at fortnightly intervals, so that sufficient amount of fertilizers is maintained in the pond always.

The brood fishes can be stocked after preparation, liming and after fertilizing the pond.

1. Chinese hatchery

The successful spawning and hatching of IMC and Chinese carps depend on the continuous flow of water. This is the main concept of a Chinese hatchery (Fig 2.4 and 2.5). Moreover, the cost of setting up a Chinese hatchery is lower than that of the other types. This system is now considered to be highly suitable for the production of quality fish seeds in India. The duration of one operation cycle is 4 days. The hatchery should be constructed in a sloppy land.

Fig. 2.4. Chinese Hatchery

Fig. 2.5. Row of chinese hatchery in Bangladesh

There are four major components in a Chinese hatchery

(i) **Overhead tank:** The water supply system for hatchery is regulated through pipes from an overhead tank. The floor of the tank should be 2.6 m above the ground level. The inner dimensions is about 5.5×2.7×2.5 m having a capacity of 30,000 litre water supply to the over head tank providing a shower type system is to be arranged by pumping water from an open well or deep tube well (Fig 2.6 and Fig 2.7). This type of shower system will provide adequate aeration to the water that is supplied to the hatchery. This type of system is provided in the carp hatchery in Bangladesh.

Fig. 2.6. Shower in Over Head Tank

Fig. 2.7. Shower in overhead tank for carp hatchery

(ii) **Spawning pool:** Spawning pool is a circular or rectangular brick masonry concrete pond. The circular tanks have inner diameter of 8 m with the total capacity of 50 cubic meters. In the centre, an outlet pipe of 10 cm dia is fitted through which, on opening the valve, fertilized eggs along with water are transferred to the hatching pool. The water supply to the pool is from the overhead water tank by means of 5 cm dia. The water supply line is laid along the outside of the wall and the inlet to the pond is provided at 14 to 16 place equally spaced and fixed at an angle of 45° to the radius of the tanks with nozzle mouth, all in one direction. The brooders consisting of 80 kg males are left in the pool and the valves are opened to create water circulation in the pool. The speed of water circulation is maintained at 30 L/min. A rectangular spawning pool is given in the Figure 2.8.

(iii) **Incubation pool:** There may be two or more hatching pools depending on the frequency of operation of spawning pool. Each hatching pool consists of 2 chambers. The outer dimension is 3 m having masonry or concrete wall. Another circular wall with fixed

Fig. 2.8. Spawning pool   **Fig. 2.9 Incubation Pool**

nylon screen is provided at 0.5 m clear distance from the outer wall. The inner chamber is provided with 5 cm dia vertical outlet with holes at different heights for taking excess water (Fig. 2.9). The spawn along with water flows from these pools to spawn receiving pond. The water flow rate is maintained at 2.5 L/sec at first and reduced to 2 L/sec when there is movement observed in the yolk. The flow rate is then increased to 3.0 – 3.5 L/sec after hatching, and must be maintained till the complete absorption of the yolk.

(iv) **Spawn receiving tank:** This tank is rectangular in shape with the dimensions of 4x2.5x1.2m. It receives water from the overhead tank through 7.5 cm dia pipe. The spawn are collected in the nets fixed to both sides of the tank.

3

TRANSPORTATION OF FISH BROODERS AND FISH SEEDS

The transportation of the brooders from the grow out farm ponds to the hatcheries and the seeds from the hatcheries to the farms is unavoidable in fish culture. Moreover, since the market demand of live fishes is more, the harvested fishes must be transported live in good conditions to the markets. Nowadays consumers prefer to buy only live fishes sold in the fish market. The various methods of transportation of the brooders and the fish seeds are given below.

(i) Transportation of brooder

Utmost care is required for the transportation of the brooders to the hatcheries. The brooders are collected in the early morning. The selected brooders must be acclimatized in cement tanks or small hapas for at least six hours without feeding. Provision of running water environment or use of showers is recommended for acclimatization. During evening, when the temperature falls, the brooders can be transported accordingly in used kerosene barrels or 500 L FRP tanks containing water oxygenated with oxygen cylinder.

Since catfishes are air-breathing in nature due to the presence of accessory respiratory organs, the oxygen deficiency would not be a major problem. Hence, lesser quantity of water is sufficient for their transportation. They can also be transported in plastic drums/buckets with some aquatic plants like *Eicchornia*. Since catfishes avoid exposure to sunlight, broken mud pots or PVC Pipes can be provided as hiding places. It must be noted that the fishes must be transported in the same water in which they are cultured or collected.

Railways or trucks can be used for long distance transport of fishes. They can be given a dip treatment with formalin solution or salt solution before transportation. Similarly optimum density should be maintained during transportation. At higher density, injuries can occur which pave the way for bacterial infection. Since the brooders appear very lethargic and tired during transportation they are more prone to attack by opportunistic pathogens. Hence, care must be taken during transportation of brooders.

(ii) Transportation of fish seeds

Fish seeds must be transported from the hatcheries to the grow out ponds for stocking. For transportation, the seeds must be acclimatized in running water (Fig 3.1) without feeding during early morning hours. Before transportation the seeds should be disinfected with 1% NaCl for 1 min or 1ppm Acriflavin bath for 30 min should be given. The seeds should be conditioned for 3 – 6 hrs before transportation. Then the seed can be transported in polythene bags containing water to one third of its quantity. A tube from the oxygen cylinder is then introduced into the bag and the oxygen is then drawn from the cylinder through the tube in such a way that its forms bubbles through the water to prevent oxygen deficiency. When two – thirds of the bag is inflated, the string is tied round the top of the bag and is ready for transportation (Fig.3.2). After reaching the destination, the seeds should be acclimatized.

Fig. 3.1. Fish seed conditioning

Fig. 3.2. Fish seed packing

Table 3.1. Number of Indian Major Carp seeds to be packed for 12 hours transportation.

Length of the seeds(cm)	Number of seeds to be transported	Average number of seeds to be transported
1	1000-10000	5500
2	500-5000	2200
3	200-1000	600
4	200-500	330
5	75-300	225
6	50-200	80
7	25-100	70
8	25-50	40

This method is most suitable for spawn or young seeds. The stocking density must varies depending upon the size of the seeds and the distance to be transported. It is better to reoxygenate the bags when the transportation time exceeds 12 hrs. But no water exchange should be done since the seeds may die due to changes in water quality parameters. The seeds will appear very much tired and avoid feeding for two days during transportation after which they return to the normal metabolism.

(iii) Stocking of fish seeds

Early morning and evening hours are most suitable for the release/ stocking of fish seeds. The seeds must be given a dip treatment in 1% salt solution for one minute or 1ppm acriflavin solution for 30 min before stocking in order to safeguard them from pathogenic infections. The seeds from the hatcheries must not be stocked directly into the ponds. The closed polythene bags containing seeds are left floating on the water surface for few minutes to simulate the inside water temperature on par with that of nursery pond. After temperature acclimatization, the bags are opened for gradual addition of pond water followed by dipping the mouth of the bag to facilitate the spawn to swim out into the pond. The seeds are acclimatized by gradual addition of pond water into it. This procedure takes a minimum of 3 hrs and it helps to reduce the mortality rate of the seeds (Fig 3.3 and 3.4).

Fig. 3.3. Fish seed before stocking

Fig. 3.4. Fish seed stocking

4

MAINTENANCE AND HANDLING OF BROODERS

Fishes like Indian major carps usually breed in natural water bodies like rivers and lakes and not in the culture pond. However, they can be made to breed as per the need throughout the year by induced breeding as well as by photoperiod manipulation or hypophysation method. Moreover, with the selection and usage of potential brooders, good quality fish seeds can be produced by the use of genetic management, cryobanking, brood diet manipulation, etc.

1. Selection of brooders

Utmost care must be given in the selection of brooders for induced breeding. The age and the sexual maturity depends on the culture period and the feed. Usually, the brooders from natural water bodies are not preferred due to the lack of information on their age and quality. Hence, it is better to select brooders from the culture pond-which are well acclimatized to pond conditions. Further, such brooders would have full record of their age, growth and their pedigree chart. Inbreeding should be strictly avoided. The brooders should not be selected from the same place. The male brooders can be procured from one farm and the females from the other. Fishes obtained from the same parents should not be crossed with each other or with their parents which would be like breeding father-daughter, mother-son and brother-sister.

Before the onset of breeding season during summer the brooders can be segregated and grown separately. In such conditions, they must be cultured at reduced density, maintaining good water quality. They must be fed with quality feed for the better development of oocyte and milt. The brooders must be screened for any infections or disease symptoms before selecting them for breeding.

Best management practices and maintenance of brooders is highly essential for quality seed production. The brooders attain ripeness or maturity of reproductive organs continuously during the breeding season and must be controlled. But throughout the year, the breeding season can also be extended. The male and female brooders should be maintained in the same pond at least during the breeding cycle period, however they can be separated during spawning, since the continuous mating and chasing by the males can cause physical injury to the brooders.

The sexual maturity of fishes depends on many external factors and with the careful maintenance of these factors, the breeding and maturity can be controlled. It has been proved that the management of photoperiod, an environmental factor can help control maturity in temperate species. However, in tropical species, the maturity does not vary much with the external factors and is difficult to identify.

The sexual maturity is controlled by the external factors like photoperiod, temperature, water and atmospheric conditions which inturn influence the secretion of pituitary gonadotropin hormone. The need for light varies with the species. Other environmental factors like rainfall, flood and water flow also influence the breeding. Due to the sudden fall in the electrolyte level in the water due to rainfall, the sexual maturity is induced. Hence, many farmers have gained success in breeding by controlling the environmental factors in carp culture.

The maintenance of brood stock can be improved by considering various factors by pond size, depth, pond preparation, right stocking density, the ratio of species, usage of healthy food, water quality improvement, aeration, etc.

The brooder ponds shall be 2000 to 5000m^2 in size less than 25 mt in width with the average depth of 1.0 meter. It is better to have rectangular ponds which should be suitable for frequent water exchange and free from aquatic weeds, predators and weed fishes. In the pond with the brood stock density of 1500 kg/ha, the ratio of different species (on weight basis) shall be Catla 450 kg, Rohu 300 kg, Mrigal 300 kg, Silver carp 150 kg and Grass carp 300 kg for effective growth and maturity. Prior to breeding season male and female can be reared separately with specific diet for effective breeding.

Provision of formulated feed with 30% protein, 20% carbohydrates, 7% lipids, phosphorus, calcium, vitamins and minerals, maintenance of oxygen balance (above 3 ppm) by aeration at night and regular water exchange once in a fortnight are the key factors for 20-25% increased growth of brooders. In India, the brooders in north-eastern states attain maturity earlier than those of the eastern states. Usually, the carps breed during February to March in West Bengal and Tripura whereas it is during April in Odisha.

In carps, the breeding can be fastened by administrating hormone (Inducing agent) injections containing Human Chorionic Gonadotrophic hormone (HCG), pituitary hormone or synthetic hormones, (Ovaprim, Ovatide, Ovopel, Wova-FH). In Indian Major Carps, the second breeding occurs within 38-73 days of the first spawning. And now, the breeding period has been extended by the

CIFA (Central Institute of Freshwater Aquaculture) to third and fourth spawning. In fishes, Vitamin C increases fecundity and make breeding easier.

2. Maintenance of brooders during summer

In the breeding of carps, the environmental temperature plays a major role. The carps usually breed naturally in the monsoon season. For the induced breeding, there is a need for oxygen rich freshwater at lesser temperatures. Hence, the carp hatcheries are set up in place with suitable environmental temperature or in places where the arrangement would be made. However, the carps which are carefully maintained are prone to suffer by increased temperature or heat during summer.The pond water temperature rises to about 38°C in the surface and 35-36°C at the bottom of the pond. This increased temperature cause heat shock and may even lead to the sudden death of the fishes. Though the increased temperature during summer do not result in death in, silver carp and common carp, it may cause indirect effects like weight loss, atresia of eggs, discolouration, skin ulcers, bacterial and fungal infections, fin rot and mortality of immunosuppressed and weak fishes. The above said symptoms are common in Indian and Chinese carps during peak summer season. Until a few years back, they were controlled by reducing the light penetration either by increasing the depth of the pond or by increasing the turbidity by plankton by fertilizing the pond. But with the research on the conditions of the broodstock and the pond environment, the brooders can be easily saved from heat shock during summer even while maintaining water at reduced depth by applying the following management practices.

- The stocking density of the brooders should be reduced to 1kg per 10m² so that the brooders will have increased unit space in the pond.

- A garden sprinkler can be set up at the middle or at both sides of the pond, which would suck the bottom pond water and spray it on the surface of the pond. This would reduce the water temperature and aerate the pond water. It also causes a gentle water current in the pond, thus providing a good environment for the brooders.

- A paddle wheel aerator/oxygen diffuser of 1 HP capacity can be used to oxygenate the pond water which also reduces the pond water temperature. A healthy pond environment can be maintained by operating it during the night time.

- The antibiotic oxytetracycline can be incorporated in the fish feed at the rate of 50mg/kg. It serves as a prophylactic measure against skin ulcers and fin rot.

- Necessary care must be given to the quantity and quality of the feed provided to the brooders. The brooders must be at the right time at the right place in required amount to meet out its needs and should not be overfed or underfed. The metabolic rates of the fishes increase during summer and hence the quantity of feed required increases. Hence, the normal quantity of feed supplied (2% of the body weight) would not be sufficient to meet out all the needs of the body and the brooders lose their body weight. So, the feed for the brooders must contain moderate amount of protein (20-25%) and increased amount of carbohydrate (5-10%). The supplementary feed given must be about 2% of the total body weight of the brooder. The total amount of feed for a day can be given in three to four batches at different times, thus preventing the wastage of feed and also helps in complete assimilation of the feed by the fishes. The production of live feed in the pond can also be improved by moderate fertilization. But at the same time it is essential to prevent the growth of blue green algae.

By following the above said practices and careful management, the hatchery owners and farmers can easily maintain their valuable brood stock and at the same time can produce quality fish seeds in larger quantity.

5

INDUCED BREEDING OF CARPS

Quality seeds are the prerequisite for successful fish culture. Though the Indian Major Carps (Catla, Rohu, Mrigal and Calbasu) and Chinese carps grow rapidly and attain sexual maturity in ponds, they do not breed in confined water like ponds. They breed in flooded rivers during monsoon. Hence, the fish farmers need to relay on these natural water bodies for seeds. In order to induce breeding, the fishes must be injected with pituitary extract or other synthetic hormones. This method of induced breeding was successfully carried out in 1957 in Indian Major Carps, 1959 in silver carp and in 1962 in grass carp in India.

Before the introduction of this induced breeding technique, the rivers were the only source of fish seeds. Though the seeds from the rivers are of good quality, they are of mixed species. Further, the seeds are available only during the monsoon period.

1. Uses of induced breeding technique

- Seeds of the single species can be obtained, without any mixing.
- Need not relay on monsoon for breeding.
- Availability of seeds throughout the year.

- Seeds can easily acclimatize to the pond conditions and farm environment.

- Seeds can easily take up the artificial and supplementary feed.

- It can help in the production of genetically improved fish species.

- Due to the usage of matured broods, increased survival rate can be obtained.

2. Maturity stages of carps

There are 7 maturity stages in the development of the gametes in carps. During the third stage, there is faster growth of the ova. The total weight of the ova would be about 20%-30% of the total body weight during the fourth stage. The fecundity of the ripe females is 1.6-2.0 lakhs in catla, 3.5-4.0 lakhs in rohu and 1.3-1.8 lakhs in mrigal.

A well matured milt of 1 kg body mass would yield 1-2 ml milt during its peak breeding season. Prospective males can yield about 6-10 ml milt on hormonal induction with any inducing agent. The quality of the milt can be identified by its viscosity, sperm density and motility.

Table 5:1. Sexuality in carps

Sl.No	Females	Males
1.	The pectoral fin will be small and thin	The pectoral fin will be long and well developed.
2.	The dorsal surface (inner side) of the pectoral fin will be soft and smooth	The dorsal surface of the pectoral fin will be rough
3.	The ripe female has a soft and bulged abdomen	The base of the abdomen will be rough.
4.	The vent will be swollen and reddish	The vent will be sunken.
5.	On gentle pressing of the abdomen, the ova ooze out.	In ripe males, milt oozes out freely when the abdomen is gently pressed.
6.	The body would appear round and bulging.	The body would appear lean and straight.

3. Raising of brooders

For a successful induced breeding programme the brooders must be raised at the required quantity till their maturity and utmost care must be given 4-5 months prior to the breeding season. Rectangular ponds having an area of 0.2-0.5 ha with average water depth of 1.5 m is preferred for brood raising. Density of the brood in the raising pond is kept at 1000-2000 kg/ha with each brood weighing about 2-4 kg. Broods of 2-3 years age are preferred (rohu and mrigal female attain maturity at the end of second year) whereas catla females attain full maturity at the end of third year. The brooder ponds must be fertilized with lime, organic and inorganic fertilizers. Conventional supplementary feed such as ground nut oil cake and rice bran (1:1) may be given. Water exchange must be done once in two weeks.

Indian Major Carps and Chinese carps can be grown together but catla and silver carp should not be stocked together in the same pond. Usually, the male and female brooders (Fig 5.1) are stocked and grown separately. Grass carp must be fed with aquatic plants like Najas, Hydrilla and Lemna at 20-25% of their total body mass. The health condition and the maturity of the brooders must be checked at least once in a month.

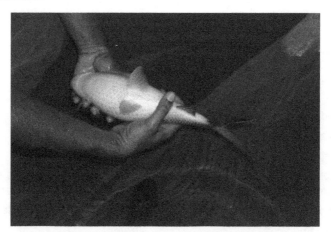

Fig. 5.1. Fully matured female carp

Selection of breeders

The selection of healthy breeders is the prerequisite for successful breeding. The development of roughness on the dorsal surface (inner side) of the pectoral fin during breeding season in males and the smoothness of the pectoral fin in females would distinguish males from females. Further, in ripe males, milt oozes freely when the abdomen is gently pressed. The ripe female has a soft and bulged abdomen and a swollen reddish vent.

4. Basic principles of induced breeding technique

Many fish exhibit reproductive dysfunctions when reared in captivity. Females fail to undergo FOM (Final Oocyte Maturation), and then ovulation and spawning. Male produce small volumes of milt, or milt of low quality. The lack of FOM, ovulation and spawning in cultured fish results from the fact that captive broodstocks do not experience the environmental conditions of the spawning grounds. Many fish migrate hundreds to thousands of kilometers to reach the environmental niches where conditions are optimal for their survival of their offspring. During this migration, the fish gained some other experience in multiple environmental changes, e.g. water salinity, chemistry, temperature, depth and substrate. These combined changes trigger the endocrine processes that result in FOM and ovulation, leading to successful spawning. In the absence of such environmental triggers, captive broodstocks become arrested in advanced stages of vitellogenesis, followed by follicular atresia. Thus, manipulation of various environmental parameters such as temperature, photoperiod, salinity, tank volume and depth, substrate vegetation, etc., can often improve the reliability of spawning.

Hypophysation

Hormonal treatments are the only means of controlling reproduction. Crude use of ground pituitaries from mature fish containing gonadotropin (GTH) which were injected into broodstock to induce spawning. Today, various synthetic highly potent agonists of the gonadotropin – releasing hormone (GnRHa) are available.

1. Crude carp pituitary

Pituitaries of captive broodstock contain the hormones required to trigger spawning. In captive fish failing to ovulate, the hormone responsible for triggering FOM and ovulation i.e., Luteinizing Hormone (LH)- is produced and accumulates in the pituitary but is not released into the bloodstream. Consequently LH does not reach the target organ, the ovaries, to trigger FOM. The first report on the ability of exogenous hormones to induce FOM and ovulation in fish was made by Houssay (1930). The use of ground pituitaries and pituitary extracts to induce spawning in fish, started in the late 1930s in Brazil followed by USA and Japan. Collection of pituitaries for hypophysation was done from reproductively mature broodstock either males or females. Pituitaries collected during the spawning season were more efficacious in inducing spawning. This was later shown to be the result of increased accumulation of GtH (primarily LH) in the pituitary prior to, and during, the spawning season. Pituitaries are preserved in alcohol or dehydrated in acetone. Ground in physiological saline prior to injection to the recipient fish. Pituitary from a donor fish for a recipient fish of similar weight was used in the case of males. Ratio was 1.5: 1 in the case of females. Priming dose 10-20% and resolving dose is given after 12-24 hrs.

The secretions of the pituitary gland are essential in the maturity and ovulation of fishes. This gland secretes two major hormones namely, the gonadotrophic hormone – 1 (GTH-I) during vitellogenesis and Gonadotrophic hormone – II (GtH-II) during the final maturation and ovulation. The amount of gonadotrophic hormone in the pituitary gland of grass carp and silver carp is about 160-200 µg. However, it depends on the amino acid sequence in fishes. The pituitary glands must be collected from matured brooders, usually of carps and catfishes. The pituitary gland is located on the ventral side of the brain. The glands are obtained prior to spawning season from matured and freshly killed fish; these can also be obtained from fish preserved in ice for 10 days. As soon as collected, the glands must be stored in refrigerated condition in acetone or ethanol.

Preparation of pituitary extract

- To prepare the extracts, required quantities of alcohol preserved glands are dried in a filter paper for 15 mins.

- The gland is weighed accurately in a chemical balance as per the requirement. Usually, the average weight of carp pituitary is about 5 mg.

- The gland is macerated with 0.9% NaCl solution.

- The prepared solution is centrifuged at higher rpm for 2-3 min and the clear supernatant extract, neither thick nor watery, is taken for use.

For the preparation of 0.1 ml extract, 2-3 mg pituitary gland is sufficient. The maximum and minimum amount of pituitary extract administered to the fishes is 1.0 ml and 0.1 ml respectively.

The exact dose of pituitary extract depends on the stage of sexual maturity and the water temperature. The female is given two doses viz, a preparatory or priming dose 2-4 mg/kg of its body weight and a second or resolving dose of 5-10 mg/kg body weight. The males on the other hand are given only a single dose. In silver carps and grass carp, the dosage is increased to 10-14 mg/ kg of body weight for females and 3-4 mg/kg of body weight for males (Table 5.2).

Table 5.2. Dosage of pituitary extract for carps

Species	Females (ml/kg)		Males (ml/kg)
	Priming dose	Secondary dose	
Catla	2-3	6-8	2-3
Rohu	2-3	6-8	2-3
Mrigal	2-3	6-8	2-3
Silver carp	4-6	6-10	4-6
Grass carp	4-6	6-10	4-6

2. Ovaprim and Ovatide

Ovaprim utilizes the fish's own endocrine system to safely induce maturation and coordinate spawning dates. When used in the normal spawning cycle, ovaprim can synchronize and coordinate maturation in treated fish by significantly advancing maturation without affecting viability or fecundity. Ovaprim has been tested and proven effective in 6 species of salmonids and several other cultured species. Ovaprim is a licensed and registered in several countries. A single dose of ovaprim is normally sufficient to induce maturation. Initial priming doses can be used to potentiate maturational effects. Injections of ovaprim are delivered to the peritoneal (abdominal) cavity using a standard needle and syringe.

In natural spawning, nerve cells in the brain deliver GnRH to the pituitary. The GnRH serves to liberate gonadotropins from pituitary cells. The gonadotropins then direct maturation of the gonads through gonadal steroid hormones. Ovaprim contains an analogue of salmon GnRH and a brain neurotransmitter (dopamine) inhibitor. The GnRH in ovaprim elicits the release of stored gonadotropins from the pituitary. The dopamine inhibitor serves to remove other inhibition of GnRH release. The outcome of using ovaprim is a burst of maturational hormones from the pituitary that induces final maturation of the gametes via endogenous gonadal steroid hormones. These steroid hormones are essential to final gamete maturation. The final maturation of gametes using ovaprim does not interfere with spawning behaviour or gamete viability. Ovaprim can be used on all teleost fish including: salmon, catfish, perch and perch-like fish, groundfish such as flounder, halibut, cod and sablefish. Ovaprim can also be used on fish such as sturgeon, gars, bowfin and others. Ovaprim is a combination of an analogue of Salmon Gonadotropin Releasing Hormone (sGnRHa), (20 mg/ml) and Domperidone 10 mg/ml, a dopamine antagonist in a stable solution. Ovaprim can be stored at ambient temperature, even in the tropics, for more than a year. Packed in 10 ml and 100 ml vials, it is the first ready to use product for inducing ovulation or spermiation in fish. As the first product available combining the hormone (sGnRH-a) and the dopamine antagonist, it has been demonstrated to be effective in a variety of freshwater and marine fishes. The minimum post-spawning mortality of the used spawners and non-adverse effect on the growth of the spawn are the merits of this drug. The Ovaprim is sold in India at Rs.490/- per 10 ml vial by Glaxo Co.

Since ovaprim is a monopoly item of the Canadian company, Syndel laboratories Ltd., Vancouver, it is available to the Indian fish breeders at a very high cost. This prompted HemmoPharma, Mumbai to undertake the synthesis of sGnRH analogs and explore the possibility of developing an indigenous and cost-effective formulation, which can be made available to the Indian fish farmer at an affordable price. Since HemmoPharma was already producing pharmaceutical peptides on large scale, they could easily develop the technology for the synthesis of active analogs of sGnRH and came up with the totally indigenous product Ovatide.

Although the biologically active ingredients of Ovatide are nearly the same as that of Ovaprim, the mixture of solvents used as the vehicle is different. This is possibly the reason that Ovatide is easier to inject into the fish. Intensive field trials with Ovatide were conducted under different agro-climatic conditions and the results have been extremely encouraging. Ovatide, an indigenous product is available to Indian farmers in approximately half the cost of ovaprim.

The dosage of various hormones is given in the following Tables.

Table 5.3 Dosage of Ovatide for carps

Species	Female (ml/kg)	Male (ml/kg)
Catla	0.4-0.5	0.2-0.3
Rohu	0.2-0.4	0.1-0.2
Mrigal	0.2-0.4	0.1-0.2
Silver carp	0.4-0.5	0.2-0.25
Grass carp	0.4-0.5	0.2-0.25

Table 5.4 Dosage of Ovaprim for carps

Species	Female (ml/kg)	Male (ml/kg)
Catla	0.4-0.5	
Rohu	0.3-0.4	
Mrigal	0.25-0.3	01.-0.2
Silver carp	0.4-0.8	
Grass carp	0.4-0.8	

Note: Minor adjustments in the dose can be made depending on the prevailing weather conditions and maturity status of the brood fish.

3. Human Chorionic Gonadotropin

The placental hormone which is glycoprotein in nature, has got ability to induce breeding and causes release of spermatozoa and ova from the mature gonads of the fish. Probably it acts synergetically with the circulating gonadotropins of the pituitary origin. HCG can be obtained from the urine of pregnant women, so, it is very cheap and can be made easily available. Commercial HCG is also available under different trade names like PREGNYL PROFASI and SUMAACH with strength of the hormone superscribed on the packet. Sometimes, the hormone (powder) and solvent are given in separate vials which have to be mixed at the time of administration as per requirement. In case the HCG powder alone is supplied, it can be mixed either with distilled water or physiological saline. Human chorionic gonadotropin is essential for the development of eggs and the ovulation and is used in many fish farms. However, many farms relay on the gonadotropin in fishes and the dosage depends on the species. The HCG hormone is a type of polypeptide hormone and its working is similar that of the Luteinizing Hormone (LH) and Follicle Stimulating Hormone (FSH).

Availability of pituitary remained restricted and the cost is very high. HCG is available in clinical grade preparation throughout the world and potency of all preparation is calibrated according to International standards. HCG given in single dose 100-4000 IU/ kg of body weight.

Method of preparation

The required quantity of hormone depending on the weight of the brooder is taken and mixed with 0.7% salt solution and can be used.

Drawbacks

Gonadotrophins are large peptides and when fish are treated with heterologous piscine LHs or HCG they may develop antibodies against them. When the same treatment is applied in sub-sequent years, the fish develop an immune response injected GtH is immuno-neutralized. Result may be that significantly higher doses are required to induce spawning or that the treatment is completely ineffective, necessitating the early retirement of otherwise productive broodstock. HCG is preferred over GnRHa.

Advantages of HCG

- It acts directly at the level of the gonad and does not require the existence of LH stores or activation of the pituitary gonadotropes.

- Occasionally, GnRHa is not effective or requires a long time to elicit a response.

- A long latency period between treatment and spawning may result in pre-spawning mortalities, due to stress induced by the capture of gravid wild broodstock, or the transport of cultured brood stock from outdoor ponds/cages to the hatchery.

- Under these circumstances, HCG may be more appropriate, because it acts much faster, in direct stimulation of the gonad, in inducing FOM, spermiation and spawning.

Types of hormone administration by injection

There are two methods for injecting hormones in fishes.

(i) Intramuscular injection

(ii) Intraperitoneal injection

Among these two, intramuscular injection is preferable

(i) **Intramuscular injection:** In this method, the hormone content is injected intramuscularly beneath the scale at the caudal peduncle region above the lateral line (Fig. 5.2).

(ii) **Intraperitoneal injection:** The hormone is injected on the thin scaleless skin below the pectoral fin (Fig 5.3). During injection, the brooders must be carefully handled to reduce stress and can be anaesthetized with MS 222 at 50-100 mg/l or quinaldine at 50-100 mg/l.

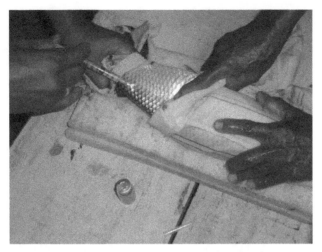

Fig. 5.2. Intramuscular injection

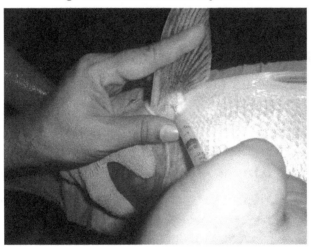

Fig. 5.3. Intraperitoneal injection

1. Examination of gonadal maturation

Well-developed eggs will be situated at the end of the ovary and can be drawn with the help of a catheter. It is an L-shaped fine polythene tube or a glass tube which is inserted into the oviduct through the vent of the female to draw a small quantity of eggs for microscopic examination. The eggs must be soaked in a solution containing 70% acetic acid and 30% pure alcohol (95% ethyl alcohol) for 5 min in a petridish. Within 5 min, the yolk would appear brighter indicating the spawning stage of fishes.

Fig. 5.4. Stripping Female Rohu

2. Artificial fertilization

With the administration of the required quantity of inducing hormones spermatogenesis, oogenesis and ovulation can be enhanced. However, the climatic conditions and the latency period (period between the hormone administration and ovulation) govern the complete maturity of the ova. The types of artificial fertilization are as follows.

(i) **Dry fertilization:** The spermatozoa are collected from the milt through gentle stripping of the abdomen of the males and maintained at 4°C. When the females become ripe for stripping (Fig 5.4), the ova is stripped into a basin by gently pressing the abdomen to which the stripped milt of the male is added immediately (Fig 5.5). The mixture is then thoroughly stirred using a quill feather to facilitate fertilization (Fig 5.6).

(ii) **Wet fertilization:** Both the milt and the ova are stripped into a basin containing water. Once the spermatozoa come in contact with water, they get activated immediately. However, the spermatozoa lose their motility in 15 to 30 sec and hence some ova may be left unfertilized.

Usually, the wet fertilization method is widely followed. But the quantity of spermatozoa collected should be accurate since the excess amount would block the micropyle of the egg. After 5 to 10 min, the remaining spermatozoa and the yolk particles must be cleared using pond water. Depending on the climatic conditions and the species, the eggs become bulged within 30 min to 2 hours. This process is called water hardening. Since the eggs of common carp and catfish are sticky in nature, the fertilized eggs get chance to clump together.

Fig. 5.5. Stripping male Rohu

Fig. 5.6. *In vitro* Fertilization

3. Methods of breeding

After the injection of the priming dose of the pituitary extract, the male and female breeders are introduced into the breeding hapa in the ratio of 2:1. Spawning occurs six hrs after the injection of secondary dose to the females.

For silver carp and grass carp, the method of artificial fertilization is followed. The ova are stripped from the females into a dry circular basin to which the stripped milt of male is immediately added. The mixture is thoroughly stirred using a quill feather and small quantity of pond water is added to facilitate fertilization. The eggs become 'water hardened' absorbing water and can be left into hatching hapa to hatch. Nowadays even Chinese carps (silver carp and grass carp) are bred like that of IMC naturally and there is no need for artificial fertilization. The reason may be the silver carp and grass carp have become domesticated.

The Indian Major Carps breed in natural water bodies during monsoon months. In India, the spawning season of Indian major carps is April to August or September and of the Chinese carps is June to September. Usually the Indian Major Carps are left in the breeding hapas immediately after injection. It is found that maintaining water temperature of 30°C or lesser increases the percentage of fertilization and hatching.

$$\text{Fertilization rate} = \frac{\text{Fertilized eggs}}{\text{Total eggs}} \times 100$$

$$\text{Hatching rate} = \frac{\text{Hatched fry}}{\text{Fertilized egg}} \times 100$$

$$\text{Survival rate (\%) of fry 4 to 5 days old} = \frac{\text{No. of stocked fry}}{\text{Fertilized eggs}} \times 100$$

Hatching: The hatching hapa comprising two rectangular enclosures is being commonly used. The inner enclosure is made from smaller meshed mosquito net cloth and the outer is made from very fine meshed nylon material. The dimensions of the hapa is 2mx1mx1m and can hold about 50,000-1,00,000 eggs. The eggs of IMC hatch out in 16-18 hours (Fig 5.8) and the hatching time is increased by two hours for silver and grass carps. The hatchlings (Fig 5.9) (Spawn) escape to the outer enclosure, through the meshes of the inner enclosure. The shells and dead eggs present in the inner enclosure are then removed.

Operation of the hatchery

Brood fishes are released in the spawning pool for about 4 to 8 hrs for conditioning. The female brooders are injected at two stages. The second dose is given 6 hrs after the first dose during which the males are also injected. Spawning occurs 4-8 hrs after the second injection depending on the stage of maturity and water temperature. Nearly 10 millions eggs can be handled during a single operation of the hatchery (Fig 5.7). The eggs are collected from the bottom and transferred into the incubation pools through pipe lines by opening the valve. Arrangements are made for water flow in the hatching pools through duck mouths in a circular motion and are maintained for 4 days till the spawn attains 6 mm size, from where they are removed into spawn receiving tanks for stocking in nursery ponds. If oxygen is less, aeration can be provided through a compressor in the hatching pool run by a 1 HP motor.

(i) Nursery rearing

Two techniques are followed in nursery rearing of fish seeds

1.   Rearing in earthen ponds in the farm

2.   Two stage rearing of seeds

(i) First stage – rearing inside the hatchery

(ii) Second stage – growing in earthen ponds in the farm

The weed fishes in the nursery ponds are eradicated by applying mahua oil cake (MOC) containing 4-6% saponin at the rate of 200-250 kg/ha. Aquatic insects are also eradicated. The pH in the pond is corrected using lime at 250-300 kg/ha. The pond is then fertilized by applying cowdung at 5000 kg/ha. The 3-4 days old spawn can then be stocked at the density of 25-30 lakhs per ha area of the pond.

Fig. 5.7. Chinese hatchery

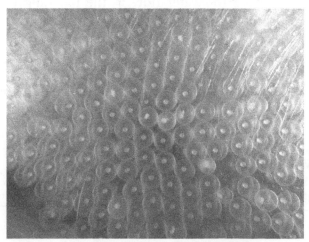

Fig. 5.8. Developing carp eggs

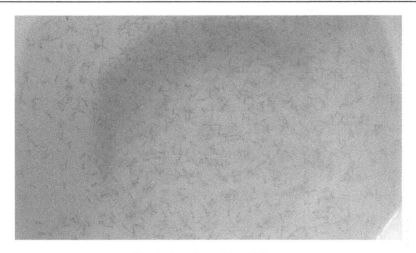

Fig. 5.9. Carp Hatchlings

Supplementary feeding

For feeding the growing spawn, a mixture of powdered groundnut oil cake or mustard oil cake and rice bran in the ratio of 1:1 is broadcasted in the pond. The feeding must be stopped a day prior to harvest. The feeding schedule is given in the Table 5.5.

Table 5.5. Feeding schedule of Indian Major Carp seeds

Days	Feeding ratio for a day	Quantity of feed for 1 lakh seeds/day
First 5 days	4 times the total initial weight of fish seed during stocking	0.56 kg.
6th – 12th days	8 times the initial weight of seed fish during stocking	1.12 kg
13th – 14th days	No feeding	–

Recently a broodstock diet CIFABROOD^{TM} is getting popularized for its effect on producing quality fish seed in case of Indian Major carps. Application of CIFABROOD^{TM} and demonstration of early gonad maturity has been achieved in the non breeding season. It has impact on breeding and breeding performances spawn recovery and fry survival of Rohu. Repeat breeding has been successful in case of spent fish and this will help to produce quality fish seed.

(ii) Harvesting of fry

Once the spawn reach 25-35 mm length with a survival rate of 70-80% they are ready for harvesting. Fine meshed seines can be used for harvesting. The harvesting of fry should be done during early hours of morning or late evening. The mortality of fry may occur during sunny periods and on cloudy days, owing to high temperature in the former case and depletion of oxygen to the latter.

(iii) Harvesting in rearing ponds

Once the fingerlings reach 100 – 150 mm after 3 months (Fig 5.11) of culture, they can be harvested using seine nets of 8.0 mm mesh size (Fig 5.10). It should be ensured that all fishes at the bottom of the pond are harvested.

Fig. 5.10. Collection of fingerlings from the nursery pond

Fig. 5.11. Catla fingerlings

(iv) Transportation of fish fry or fingerlings

Transportation of seeds to longer distance can be done in airtight containers filled with oxygen rich waters. Usually polythene or polyvinyl chloride or other plastic bags of 15-35 L capacity are used for transportation. Further details are discussed already in the Chapter 3.

6

SIGNIFICANCE OF QUALITY SEEDS IN SUSTAINABLE FISH PRODUCTION

Production of high-quality seed is fundamental to modern aquaculture. Most annual crops are established each season from seeds, and seed quality can have a major impact on potential crop yield. Seeds can serve as the delivery system not only for improved genetics but also for new stocking and production methods. Fish has a major role in the income generation of farmers and being the simple food for the poor. In the global scenario, India stands second in fish production next to china.

In aquaculture, it is essential to produce good quality fish seeds in large quantities, especially of the carps, since carp culture forms the mainstay in Indian aquaculture. Unfortunately, the lack of genetic improvement has resulted in the production of poor quality seeds in many fish hatcheries. Moreover, nearly 5-10% of the production cost is spent on the purchase of fish seeds. But lack of production of good quality seeds for the culture practices has resulted in reduced growth and production, seriously affecting the income and the economics of the fish farmers.

What are Quality fish seeds?

Fish seeds with higher growth rate, low FCR, compatibility with the changing environmental conditions and good disease resistance are quality seeds. Poor quality seeds can result in about 30% loss in total production.

Bottlenecks in quality seed production

- Lack of certification for fish seed production in India.
- Lack of government regulations for private fish hatcheries.
- Production of hybrid fishes because of the simultaneous breeding of different fish species in the single, large breeding ponds.
- Introduction of exotic fish species without any impact assessment.

Reasons for negligence of genetic principles in commercial carp hatcheries

- Lack of knowledge on the effects of continuous use of the same parental stock for breeding.
- Broodstock are not changed now and then.
- Absence of pedigree charts and records about the broodstock.

Best Management practices for Quality Seed Production

Nowadays, selection of brood stock is given lesser importance in many of the fish hatcheries. Much concern is being given on increasing profit through increase production. Priority is given only for quantity and not quality. The seeds produced are first sold to the farmers for culture and only the rest are stocked for growout in their own farm. And some of these fishes are used for breeding at the ratio of 2 males to 1 female, two months prior to their breeding season. Unfortunately, this practice is followed nearly in all the states. But, main focus must be made only on the fast growing fishes which must be selected and used for breeding.

It is better to grow the brood stock in composite fish culture ponds. The ponds must be rich in plankton. The male and the female broodstock must be 2-4 years old and must be maintained with special care 5-6 months prior to the breeding season. Moreover, the fish seeds must be procured from quality hatcheries and farms after knowing their genetic characters.

i) Number of broodstock

In the context of quality fish seed production, the number of broodstock to be maintained in the pond gains importance. Nowadays, brooders are maintained in lesser numbers which is a wrong practice. Even when larger number is maintained, only a few are used for breeding. This method of brood stock selection must be avoided.

Every farm should maintain a minimum of 50 to 500 males and 50-500 females as brooders for each species of fish and used for breeding. The breeders selected for induced breeding must not be closely associated to facilitate increased genetic difference.

Every year during breeding season, the older broodstock must be replaced with the new ones. Maintaining a pedigree chart helps in easy selection of brood fish. Either the male or the female brooders must be exchanged among the farms.

The stocking density can be maintained at 1500-2000 kg/ha. The health of the brooders must be checked at regular intervals to prevent diseases. The ponds must be replenished with oxygen rich water 2-3 months before the breeding season and water exchange must be done once in two weeks.

Since the natural food available in the ponds will not be sufficient, artificial feed can be given 1-3% of the body weight of fishes. When feeding is done *ad libitum* and with better management practices, the fishes attain maturity before the start of monsoon. But with excess feeding, the attainment of maturity is arrested due to excess deposition of fat.

ii) Maintenance of brooders during breeding

Special attention must be given in the maintenance of brooders before and after the spawning.

- The brooders must be handled gently. They must be transported from pond to spawning tank/pond with water in buckets so that they are not stressed.

- Intramuscular injection beneath the scales is more preferred than intraperitoneal injection.

- The brooders must be maintained in water showers before and after the injection

- The brooders must be given a bath treatment in potassium permanganate solution for 5 min after spawning.

- The spent brooders must be stocked in ponds with oxygen rich water, sufficient feed and a clean environment.

iii) Maintenance of fertilized eggs and the fry

- The eggs must be collected from the spawning pool and maintained in the hatching tank.

- The water flow must be controlled so that the eggs are prevented from the inner screen and the incubation chamber. The flow rate of water is maintained at 0.4-0.5 m/sec for the first 12 hrs, 0.1-0.2 m/sec for next 6 hours and then again increased to 0.3-0.4 m/sec.

- The control of water flow rate after 12 hours of fertilization prevents the hatching of underdeveloped eggs and allows only the fully developed eggs to hatch out.

- The dissolved oxygen content of the tank should not be allowed to decrease below 4 ppm.

- The water hardness must be maintained between 60-80 ppm.

- The density of hatching tank should be maintained at 7 lakhs/m^3.

The unfertilized eggs are more prone to bacterial infections and the unfertilized eggs must not be mixed with the fertilized eggs since they may also be infected.

The hatchlings are released after the complete absorption of the yolk sac within 72 hrs, plankton must be given as feed for the larvae. The hatchlings can then be collected for stocking in nursery ponds.

iv) Prevention of inbreeding

Inbreeding occurs during the mating of close relatives and it results in genetically identical or homozygous individuals. All the fishes have some rare harmful genes which may not be expressed during heterozygous condition. But due to the mating of close relatives, these harmful genes get expressed due to the sharing of the genes which may result in reduced growth rate, survival rate and disease resistance.

Every fish hatchery manager wants to increase the number of brooders but at the same time reduce inbreeding. Maintaining different species of carps in the same breeding tank at the same time may affect the seed production and should be maintained in different tanks to retain the purity of the genes. This also helps to prevent inter specific hybridization and increase heterozygosity.

It is advisable to stock seeds of different age groups in the sex ratio of 50:50 in a pond and the same can also be followed in breeding.

Females of 3-5 years age can be bred with males of 2-5 years to prevent the loss of valuable genes by genetic drift.

Brooders from natural water bodies should also be used for breeding time to time. The hatcheries can also exchange brooders among themselves and breeding of the same species of fish from different hatcheries increases the quality of seeds. Breeding of fishes of same age group, close relatives and usage of same brooder for several consecutive years should be strictly avoided. The maintenance of a pedigree record of the brooders also help to prevent inbreeding. Use of cryopreserved sperms is recommended since there is no need to maintain male brooders in large number and also this saves the place and time.

Since all hatchery managers demand quality fish seeds, importance must be given to improve the breeding conditions in fish hatcheries.

Significance of genetically improved fish seeds

Some genetically improved fish species have contributed significantly to the fish production. They are as follows.

i) Jayanti Rohu

The Central Institute of Freshwater Aquaculture (CIFA) was the first Institute in India to introduce selective breeding in fishes. This institute in collaboration with a research organization in Norway has produced the genetically improved rohu called the "Jayanti Rohu" in 1997. They were then given to different hatchery owners and the quality seeds produced were distributed to farmers for culture which showed increased growth rate. CIFA collected rohu from five different rivers of India namely the Ganga, Yamuna, Sutlej, Brahmaputra and Gomathi and produced an improved "Rohu" through selective breeding technique. This is called "Jayanti Rohu".

Though its external appearance is the same as that of the normal species, "Jayanti Rohu" shows increased growth of 18% and it is continuing for seven generation. It is being popularized among farmers with the help of the private and government hatcheries and is being monitored by the CIFA. Till now, it is being popularized in 14 states in India. Nearly 20 million seeds of "Jayanti Rohu" are popularized every year by the CIFA alone. Farm trials are conducted now and then in Odisha, Andhra and Uttar Pradesh. The growth performance is improved continuously for generations. Following Rohu, research on selective breeding for increased weight gain in catla and scampi are in progress in CIFA.

ii) Improved freshwater prawn

The giant freshwater prawn, *Macrobrachium rosenbergii* is a potential species for culture in the global level. In 2005, India was the third in its production with about 42,800 tonnes. But, the production is showing a declining trend since 2006 and it has reduced to 3332 tonnes in 2012-13. Genetic changes in the brooders and poor quality seeds are the main reasons behind the decreased production. To solve this issue, the CIFA in collaboration with the World Fish Centre in Malaysia has started the selective breeding in 2007. *M. rosenbergii* were collected from Gujarat, Kerala and Odisha and were bred. The prawns were selected based on the weight at harvest for four generation.

iii) Minor carps and barb fishes

Polyculture of minor carps and barbs increases the production and is a profitable venture. It has been proved by the CIFA that minor carps like *Labeo calbasu (calbasu),Labeo fimbriatus (fringed lipped carp), Labeo gonius (Labeo), Labeo bata (bata), Cirrhinus reba (Reba)* and barbs like *Puntius sarana (olive barb), Puntius gonionotus (Silver barb)* can be polycultured.

The induced breeding of these minor carps and the barbs has been standardized. The pond preparation steps for these minor carps are the same as that of the Indian major carps. The seeds of IMC are grown in nursery ponds for 15-20 days till they reach 25mm in length. But, the nursery culture is extended to 25-30 days for the minor carps till they reach 15 to 20mm.

In fish farms, the new technique of culturing IMC in polyculture ponds and minor carps in seasonal ponds has been introduced. Polyculture of Minor carps and barbs with the Indian Major carps increase the yield by 30%. Since the growth of these minor carps is same as that for the major carps for the first 3-4 months, they can be grown as an intermediate crop in seasonal ponds for a short period.

Benefits of culturing minor carps and barbs

- Minor carps and barbs are sold at a higher price than the major carps and also have consumer preference.
- They reach the harvestable size within 6 months.
- They can be polycultured with major carps and will not affect their growth.
- Culture of minor carps and barbs increases the production and yields more profit to the farm.
- They can survive in seasonal ponds even at lower water level.
- Their breeding and seed production is easier.

Economic importance of quality seed production

The economic conditions play a major role in quality seed production. This does not necessities major alternations in the pond since all the seeds are sold at the same price in the market. Hence, the government has brought out technique to encourage the production of quality fish seed production.

7

DO'S AND DON'TS IN THE PRODUCTION OF FISH SEEDS

In India, carp culture accounts for more than 80% of production in freshwater aquaculture. The main reason for this is induced carp spawning was successful during late 1950's following that, during post-independence period many steps were taken by the Government of India and more areas were brought under aquaculture to increase fish production. Since fish seeds were not sufficient from the wild or natural water bodies, more fish hatchery was established in all the states to increase the farming area. Following this, the need for fish seeds was increased. So, farmers were made to depend on fish hatchery. Due to these practices, carp production increased to the maximum.

The fish hatchery owner's primary aim is to produce large quantity of fish seeds, but not the quality. They didn't follow genetic procedures for fish seed production. If the fish seed is produced in hatchery, then same fish seeds will be used for breeding. This method is followed for more than 50 years. Because of this wrong practice, carp fish culture has been totally changed. Now the fish farmers are following two years of culture instead of one year culture period, i.e., stunting the seeds during first year and rearing in the second year in grow out systems.

This type of farming practice is being followed in Andhra Pradesh (Kakinada, Godavari & Krishna districts) and also in Thanjavur and Thiruvarur districts of Tamil Nadu. The stunted fishes grow bigger and faster in the grow out ponds and catla can grow up to 1.5 kg within 6 months period.

Inbreeding depression

If we study the breeding practices of IMC followed in our country, inbreeding has been followed in most of the fish hatcheries. The same brooders were used for many years and also the brooders were selected from the same hatchery where brother – sister mating is possible. This type of method will lead to inbreeding depression. It will affect the growth and reproductive activities. That means less number of eggs will be produced, reduced fertilization, hatching rate, etc. Inbreeding depression was found to occur more in many carp fish hatcheries across the country. They found that growth was affected to 10% in catla, 6-12% in rohu, 4-12% in mrigal, 10-17% in grass carp, 4-13% in silver carp due to inbreeding.

Ways to avoid inbreeding depression

- To produce fish seeds, good quality brooders should be used. Brooders should have different genetic characters.

- Brooders should be handled with 1:1 male and female sex ratio. Also male and female fishes should have same age. From 50 to 100 separate male and female brood fishes for different species should be maintained in each hatchery.

- In the hatchery, brooders should be changed for every three years in rotation. It should not be used again and again for breeding.

- If any large size brooder was caught from natural water also it should not be used for breeding because of the unknown age.

- Brooders can also be selected from different hatchery from different locations for increasing genetic variation.

Selective breeding

Selective breeding technique is the method of choice for increasing growth in fishes. It aims to produce genetically higher quality cross bred fishes. Selective breeding technique has been followed in fishes such as common carp, Atlantic salmon, tilapia, rohu, rainbow trout, etc.

Selective breeding techniques in fish seed production

Domestication

Domestication is the most important method having impact on genetic diversity. Induced breeding in an artificial or controlled environment and arresting of breeding in an unconditioned environment are the important factor in domestication. Domestication was mainly carried out in fishes such as common carps and Atlantic salmon. The domesticated common carps will be different from undomesticated common carps by some characters like body appearance, reproductive character, body growth and habitat. The Indian major carps are also highly domesticated.

Breeding by Artificial selection

In fishes breeding by artificial selection method was introduced by Norway scientists. They carried out the research for 15 years (1975-90) in salmon and trout and increased growth of 60-70% was observed. Similarly, research was conducted in Nile tilapia in Philippines (GIFT – Genetically Improved Farmed Tilapia) through selective breeding technique. Similarly in India scientist had developed fast growing rohu through the same technique and named it as "Jayanti Rohu".

There are different genetic improvement techniques available such as production of sterile fishes using chromosome manipulation technique, production of mono sex fishes using sex reversal. Although different genetic techniques are available, only selective breeding technique has proved as the best technique for the improvement of different trait in fishes. Farmers can also adopt this selective breeding technique in the field effectively by selecting suitable fish species.

Selective breeding can be done by using following techniques.

1. Individual Selection

2. Family selection

3. Within family selection

4. Cross breeding

1. **Individual Selection:** It is one of the simplest and easiest method. In this method, fishes should be separated based on their physical appearances such as fast growth, increased weight, disease resistant, etc.

2. **Family selection:** In this method, the sex ratio is one male and one female in the ratio of 1:1. The growth rate is compared among the different families and the families which showed higher growth rate can be selected, others can be eliminated or culled. The parents can be selected by studying the performance of the offspring. This method is being followed in fishes such as rainbow trout and Atlantic salmon.

3. **Within family selection:** In this method, best performing individuals will be selected within the family. In Philippines, the World Fish Centre has produced Nile tilapia exhibiting 85% increased growth rate by means of family selection and within family selection. Thus, GIFT strain was developed.

4. **Cross Breeding:** Among the selective breeding methods, cross breeding is a successful method to increase the genetic diversity. Crossbred species will have many superior characters than the parental species. Cross breeding is the opposite of inbreeding. This method has been practised in fishes such as common carp and catfishes.

8

CIFABROOD™ FEED AS A HEALTHY FEED FOR BROOD FISHES

This CIFABROOD™ feed was developed by ICAR- (Central Institute of Freshwater Aquaculture) in Bhubaneswar for the purpose of development of quality broodstock. This feed was given to carps as an experimental feed by the farmers for 2 years in their hatchery. The experimental result obtained by the farmers was very much encouraging and interesting.

Why this CIFABROOD™ is special feed for brooders

The CIFABROOD™ is a healthy feed for the brooders due to the following reasons.

i)   Production of fish seed with good quality

ii) High nutritious feed

iii) High growth rate

iv) Higher survival rate (70%)

Special Characters of CIFABROOD ™

i)   This feed is composed of required vitamins and minerals.

ii) Development of gonadal organs.

iii) Brood fishes can be used for multiple breeding.

iv) Production of good quality fish seeds.

v)   High growth rate and high survival rate of fish seeds.

Experimental results gained from the fish farmers by using CIFABROOD™ in their farm

i)   Limited number of brooders can be maintained thereby we can reduce the maintenance cost.

ii) Promotes early reproduction. i.e. before normal breeding season.

iii) Production of fish seeds is 100 times better.

iv) If the fish was bred for one time, it will be ready for next spawning within 45 days interval.

v)   Higher Survival rate of fish seeds, about 50-70%.

vi) The fish farmers were able to breed catla, rohu and mrigal before the monsoon season (i.e) during the summer season.

vii) This special feed, CIFABROOD ™ was commercially produced by M/s. Aishwarya Aquaculture Pvt. Ltd. Naihati, West Bengal. Also, the technical agreement was signed by CIFA with this farm.

Table 8.1. Differences between CIFABROOD™ and Normal feed

Sl.No	Normal feed	CIFABROOD™ feed
1	Fish brooders will get matured at 2-3% body weight of feed and it needs about 90-120 days to take place	Brooders will get matured at 3% of its body weight. It takes place at 40-50 days and then it extends to 30 days
2	When the normal feed is used, it is needed to maintain more number of brood fishes for each year	When CIFABROOD ™ feed is used, it is sufficient to maintain less number of brooders

[Table Contd.

Contd. Table]

Sl.No	Normal feed	CIFABROOD™ feed
3	The cost of feed spent for 1 kg brooders is Rs.30/- and for 1000 kg brooders it will be Rs.81,000/-	The cost of CIFABROOD™ feed spent for 1 kg brooders is Rs.70/- and for 1000 kg brooders it will be Rs.90,000/-
4	Maturity – 75%	Maturity – 92%
5	Fecundity:	
	i) Rohu and mrigal – 1.5 lakh/kg body weight	Greater than 1.5 lakh/kg body weight
	ii) Catla – 1 lakh/kg body weight	
6	Fertilization rate:85%	Fertilization rate:95%
7	Hatching rate:75%	Hatching rate:92%
8	Recovery of hatchlings: 65%	Recovery of hatchlings: More than 90%
9	Survival rate of small sized hatchlings to large sized hatchlings: 35%	Survival rate of small sized hatchlings to large sized hatchlings:65%
10	Survival rate of large sized hatchlings to fingerling stage : 55%	Survival rate of large sized hatchlings to fingerling stage: 75%

In the year 2018 National Fisheries Development Board, Hyderabad has funded a project to popularize the scientific aquaculture farming with the application of CIFABROOD™. At present around 21 hatcheries from 7 states are undertaking CIFABROOD™ demonstration.

9

INDUCED BREEDING AND *IN VITRO* FERTILIZATION IN COMMON CARP

Common carp brooders (Fig 9.1), which had grown about 1-2 kg can be selected from the farm for the development of broodstock. These brood fishes can be reared separately or along with other carps. However,it is best to maintain these brood fishes in separate ponds, with a stocking density of about 600 kg/acre in the brooder pond. It is good to feed the brooders with groundnut oil cake and rice

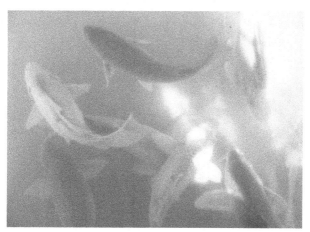

Fig. 9.1. Common carp brooders

bran at the ratio of 1:2 because it contains 28-30% protein in it. Generally, Indian Major Carps such as Catla, Rohu and Mrigal will breed in the monsoon season from June to September. At the end of the summer and before the start of monsoon common carp will be ready to spawn. Before one month of the breeding season, male and female brooders (Fig 9.2) should be stocked in separate ponds and maintained.

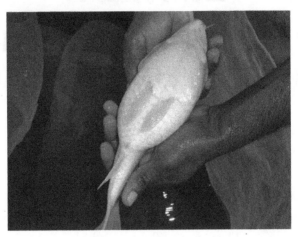

Fig. 9.2. Fully matured female common carp

Natural breeding method in common carp

In this method, male and female should be selected at the ratio of 1:1 for induced breeding technique. To breed the common carp, induced breeding hormones should be injected to the fish as per the recommended dosage and released in breeding hapa. After some time, the brooders themselves will involve in breeding due to the stimulation of hormone injection. Generally, common carp eggs have the natural ability to stick among themselves. Hence, it should be provided with nylon fibres to help to stick the eggs in it. Fertilized eggs will be yellow in colour and unfertilized eggs will be in white colour. Then these nylon fibres with eggs should be taken out and put in the Chinese hatchery. These eggs will be hatched out within 3 days depending upon the temperature.

Artificial fertilization using induced breeding method in common carp

To breed the common carp artificially, male and female fishes should be injected with inducing hormones. Males are injected at the rate of 0.2 ml/kg body weight and females are injected (Fig 9.3) at the rate of 0.3 ml/kg body weight. After injection the males and females are released in separate hapas (Fig 9.4). After 6 hrs females should be checked whether ready to release eggs. If ready, then the eggs are collected gently by stripping the abdomen (Fig 9.5) in one vessel or trough. Similarly, male fishes were checked by pressing the base of the abdomen and milt (Fig 9.6) is collected in the same vessel/trough above the eggs (Fig 9.7), at the ratio of 1:1 (male:female). It should be thoroughly mixed with the help of bird's feather (Fig 9.8). During mixing (Fig 9.9), water is gently added over the eggs- milt mixture. At the same time, 0.3% urea and 0.4% common salt (NaCl) solution is added into the container during mixing. Because of this, both Urea and NaCl will help the sperms to be active. After this mixing, the eggs will get fertilized immediately and the fertilized eggs will get attached to one by one and

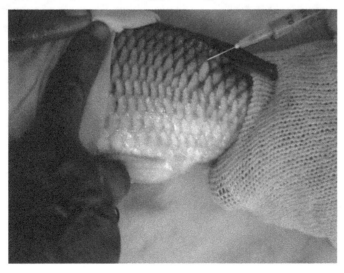

Fig. 9.3. Injection of hormone to female common carp

looked like the grapes. The eggs which was attaching outside only will get oxygen and the interior eggs may die due to the non – availability of oxygen. Hence, it is required to remove the stickiness nature from the eggs, by using milk and milk powder.

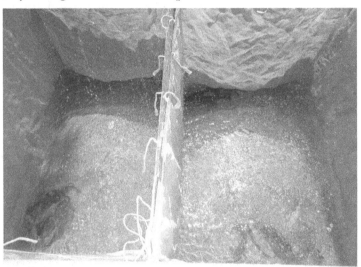

Fig. 9.4. Injected males and females released in separate haps

Fig. 9.5. Stripping female common carp

Fig. 9.6. Stripping male common carp

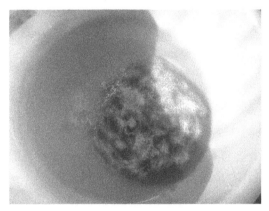

Fig. 9.7. Eggs and milt before fertilization

Fig. 9.8. *In vitro* fertilization

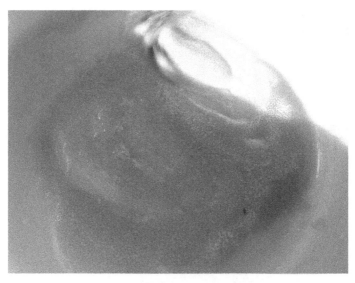

Fig. 9.9. Mixing eggs and milt

To remove the stickiness of eggs, the fertilized eggs are mixed in 200 ml milk mixed in 1 L water or 20 g milk powder mixed in 1 L water in a separate vessel. After thorough mixing of eggs, water should be drained out from the eggs. Then this milk and milk powder added water was mixed with eggs and mixed for 10 min. After 10 min duration, this water should also be filtered and drained out. To remove the stickiness in eggs, this procedure is repeated for 3 times at 10 min duration. After the removal of stickiness, the eggs should be released in Chinese hatchery with good water flow. Within 3 days, the fertilized eggs will get hatched out and 1 kg body weight of female carp can produce 1 to 1.5 lakh eggs. After that, the hatchlings (Fig 9.10) can be stocked in hapa or in the nursery pond, for rearing.

Advantages in artificial fertilization technique

By this artificial fertilization technique, we can produce many crossbred hybrid fishes and hybrids have high growth rate. When the fishes spawn naturally, all the eggs in the body will not get released out. But when it is bred artificially and *in vitro* technique more number of eggs could be obtained from the body and more number of hatchlings can be produced. Since, there is less number of bones in common carp, farmers and fish eaters are preferring it more when compared to other fishes. Thus, this technique will be more useful for the farmers to produce good quality of common carp seeds in their hatchery.

Fig. 9.10. Common carp hatchlings

How to estimate the number (kg) of broodfish of common carp to produce one million fingerling?

Average weight of females = 2 kg

Number of eggs per 1 kg of female = 1,00,000

Fertility rate of eggs 80% = 80,000

Hatching rate of fertile eggs 80% = 64,000

Survival of larvae upto 7 days 70% = 44,800

Suirvival of fingerling upto 1 month 30% = 13,400

To produce 1 million fingerling would require 1,000,000 : 13,400

= 74.6 kg (females)

Real number of females = 74.6 kg; males = 149.2 kg (with ratio 1 : 2)

Estimated resource 20% 14.9 kg; males = 29.8 kg

Females 89.5 kg; males 179.0 kg = 268.5 kg

Estimated amount of broodfish = 268.5 kg

10

FARMING OF LOACHES WITH CARPS

L oaches forms one of the most commercially important fishes among the freshwater fishes. It is distributed in India, Srilanka, Myanmar and Thailand. Generally these loaches were found in small river banks, ponds, canals and streams. Since these loaches are fast swimmers, it swims along the water current. Hence, it can be collected against the water current by using special methods. Among the freshwater fishes loaches are highly delicious and highly priced fishes. Loaches are commercially important fishes and found mainly in Madurai, Theni, Thenkasi, Tirunelveli districts of Tamil Nadu. Loach fish curry fetches the special value with highest price in restaurants. In tamil it is called "ayiraimeen" (Fig 10.1) The approximate cost of 1 kg loaches ranges from Rs.1500 to Rs.2000/-. Not only are the loaches used as edible fishes, it is also used as ornamental fishes depending upon the scarce and market demand.

The maximum length of loaches is 7.3 cm and maximum weight is 3.72g. It could be grown well in ponds with 6.5-7.5 range of pH and 22-29°C of water temperature. It breeds during April to June months. It is very difficult to differentiate the male and female fishes. Female fishes lay eggs at one month interval. It lays 1000 – 15000 eggs. Hatchlings will come out from the fertilized egg on 5th – 6th day.

Fig. 10.1. Loach (Ayiraimeen)

Although loaches possess high market value, the culture of loaches is still a puzzle. However, Mr. S. Pugazhendi, one of the Fish farm owner near Mannargudi, Thiruvarur district, is successfully culturing loaches for about more than 5 years.

Special care should be taken in loach culture

- Loaches can be grown in ponds along with catla and rohu.
- It should not be cultured along with mrigal and common carp, since they are also bottom dwellers and there will be competition for the niche.
- It can also be cultured along with carp brooders in the rearing of brood fish pond.

Similarly, the pond should be prepared same as like that prepared for carp culture. The pond should be prepared with organic manures like cowdung @ the rate of 1 ton/acre and allowed for the development of planktons. After 5 days, carp hatchlings can be stocked in the pond.

After stocking the carps, loach brooders can be stocked in the same pond. Same feed which was used for carps, like groundnut oil cake and rice bran can be fed for loaches alone. Additional feed for loaches is not required. After rearing period of 4 months, carp hatchlings grown to the fingerling size they can completely harvested. After the removal of fingerlings of carp, loaches should be harvested (Fig 10.2) after completely draining of water (Fig 10.3). During the rearing period of carps, the loaches will go on continuous spawning and grow.

Fig. 10.2. Loach harvest

Fig. 10.3. Draining of water

For example, if 2kg of loach hatchlings were stocked in 20 cent nursery pond and grown for 4 months around 30 kg of loaches can be harvested in addition to carp seeds. If the carp hatchlings were grown to the fingerling stage, then it should be harvested and stocked in growout ponds. Then the water should be drained. After complete draining of water, a pit should be constructed in the middle of the pond and water is pumped to fall out the pit. In that pit, all loaches of that pond will move to the pit, since the loaches swim against the water flow. Finally from that pit, loaches can be harvested.

If the loaches were cultured along with carp brooders means, then 10 kg of loaches can be stocked in 1 acre pond and grown for 6 months. When it is harvested after 6 months it will be more than 60-80 kg loaches. Due to the culture of loaches with carp brooders, there will not be problem in stocking density, food scarcity and water quality.

Since loach farming is gaining more popularity in Tamil Nadu, it can be grown along with carps to yield more additional income.

11

INDUCED BREEDING AND SEED PRODUCTION IN CATFISHES

The important catfish species include *Mystus seenghala, Mystus gulio, M. aor, Pangasius pangasius, Wallago attu, Clarias magur, Ompok bimaculatus* and *Heteropneustes fossilis.* Among these, catfishes belonging to the genus *Clarias* of the family Clariidae occupy important place in Asian aquaculture. The most important among the clariid fishes is the Asian catfish, *Clarias magur*, commonly called as 'magur'. The other species available in Asian markets include *C. macrocephalus* and *C. gariepinus.* The aquacultural traits of these species are attributed to good taste, fine flavour, fast growth rate, amenability to culture under high stocking density, ready acceptability to artificial feed and good market potential especially in living condition. The species *Clarias magur* is well known for its nutritive and therapeutic value. Reports indicate that the flesh of this species is rich in physiologically available iron and copper, aiding in hemoglobin synthesis and formation of blood. Apart from this, the fish contains easily digestible fat and essential amino acids, thereby making it ideally suited for sick and convalescent and also those recuperating from illness. These catfishes have high consumer preference, their breeding and culture techniques are yet to be well established in India. *Clarias batrachus* is indigenous to India and the Philippines, but are widely cultured in Thailand. In many countries of Africa and Asia, catfishes of the genus *Clarias* are commercially important.

Broodstock Management

Generally, catfish brooders (Fig 11.1) should be selected and stocked in small quantities in brooders tank. Based on our requirements, brood tanks can be constructed. Catfishes can be grown both in the cement tanks, and also in earthen ponds. But when it is grown in ponds, its collection will be very difficult but it will have good growth rate when compared with reared in cement tanks. One year old matured fishes with 150 – 200 g weight could be ideal for breeding. Male (Fig 11.2) and Female (Fig 11.3) brood fishes should be maintained in separate tanks. If the tanks had good drainage capacity then more number of fishes can be stocked. Water exchange should be done at 3 days interval. The feed which was given to brooders should be with high nutrients. It should be fed with trash fishes and rice bran at 9:1 ratio for 5-10% body weight for daily two times. If excess feed is given then it will spoil the water quality parameters, and cause diseases in fishes.

Fig. 11.1. Catfish broodstock

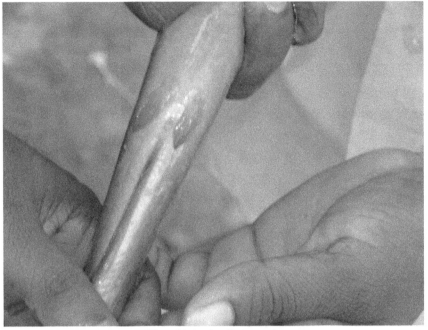

Fig. 11.2. Catfish male

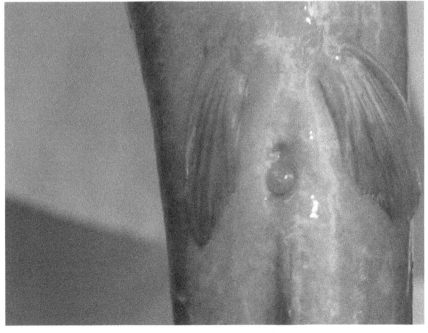

Fig. 11.3. Catfish female

Brooders can also be fed with prepared pellet feeds. Fish flour, rice bran, groundnut oil cake, wheat flour and vitamins can be mixed with water at 4:1.4:3:1.5:0.1 ratio. Initially feed can be boiled without addition of vitamins. After that with pelletizer it should be produced at 3-5 mm size. The prepared pellet feed should be dried in sunlight and packed in airtight containers. Besides these, pea flour, wheat bran, fish flour, silk worm flour, and soya beans can be mixed at 3:3:2:1:1 ratio and fed for catfishes. Between one week or two weeks interval, the boiled paste of weed fishes can be given as feed, so that it can fulfill the need of required nutrients.

Selection of brooders

For breeding good matured male and female fishes should be selected. Unlike carps, catfish males will not release milt on gentle pressure, instead the secondary sex organ (i.e.) testis will have better development and can be easily identified. Matured female brooders will have broad soft belly region, so that it can be pressed to release eggs. Such female brooders should be selected. Catfishes can be bred from May to November months, because these months water will have $21\text{-}31^0C$ temperature. Each catfishes can be bred for 6-8 times. Catfishes can be bred artificially by using different inducing hormones which is given in the following table.

Table 11.1. Different induced breeding hormones with its required concentration is given below

Sl.No	Induced breeding hormones	Female Brooders	Male Brooders
1.	Pituitary gland	5-10 mg/kg	1-1.5 mg/kg
2.	Chorionic Gonadotrophin	500-800 IU/kg	250-500 IU/kg
3.	Ovaprim	1.2-2 ml/kg	0.1-0.3ml/kg
4.	Ovatide	1.0-2.0 ml/kg	0.1-0.3 ml/kg
5.	Chorionic gonadotrophin and Pituitary gland	500 IU+1mg/kg	500 IU+0 mg/kg

Note: The IU stands for International unit. It is a unit of measurement for the amount of a substance, based on measured biological activity or biological effect.

Generally female brooders should be injected with increased amount of inducing hormone when compared with male brooders. Brooders should be handled with special care during hormone injection. Injection should be given at the base of dorsal fin (Fig 11.4). Mostly, injection should be given during evening hrs, and male and female brooders should be released in separate small tanks for overnight.

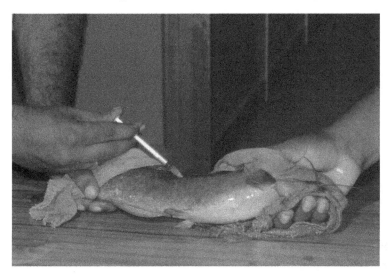

Fig. 11.4. Injection of hormone to female catfish

Artificial fertilization

After 9- 12 hrs of injection, male and female brooders will get ready for spawning. In males, abdomen region should be cut open and testis (Fig 11.7) is taken out and kept in 0.9% saline (Fig 11.8) (salt mixed with water). After that sperms can be taken from the preserved testis, when it is required. Now female catfish is taken and the eggs are stripped (Fig 11.5 and Fig 11.6) in an enamel tray. Sperms and eggs from the females can be mixed with bird's feather (Fig 11.9). After 2 min of mixing, water should be added to that mixture. During mixing, fertilized eggs (Fig 11.10) will get stick with one another. Hence there is a chance for mortality of fertilized eggs due to insufficient oxygen. Hence, fertilized eggs should be mixed with 1% sodium sulphite mixture or 20 g/L milk powder for 30 min. By this process stickiness nature of fertilized eggs will get released like that of common carp. Then the fertilized eggs are released in flow through system (Fig 11.11) with good water flow. To prevent fertilized eggs from fungal attack 1 ppm methylene blue should be added with water.

Fig. 11.5. Egg release in female catfish

Fig. 11.6. Stripping female catfish

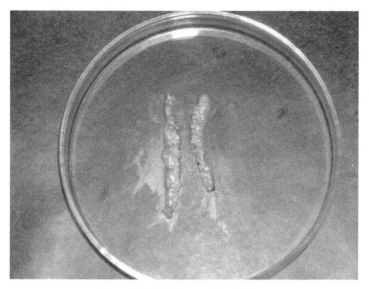

Fig. 11.7. Testes of catfish

Fig. 11.8. Catfish sperm suspension preparation

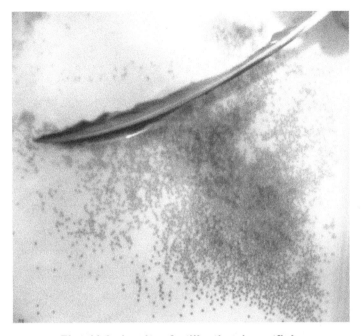

Fig. 11.9. *In vitro* fertilization in catfish

Fig. 11.10. Fertilized eggs

Fig. 11.11. Flowthrough system for incubating the fertilized eggs of catfish

Catfish hatchery needs pure water and good airflow. The water temperature should be 27-30°C. Within 24 hrs, the developing eggs (Fig 11.12) hatch out and hatchlings (Fig 11.13) will be released from the fertilized eggs. 20,000 – 30,000 eggs can be stocked in hatching tank, catfish hatchlings will be 4-5 mm length and 2.8-3.2 mg body weight. Young hatchlings will have yolk sac in it. It will get energy for 3 days by absorbing the yolk sac. When the eggs hatched out, dead eggs and egg shell should be separated from the hatchery. After 3 days, when the yolk sac is absorbed, it starts to take feed.

Rearing of hatchlings

The incubation of catfish eggs and hatching (seeds) is very difficult, so it should be done with special care. Otherwise diseases will be caused by bacteria and fungi. Hence special care is required to maintain the hatchery water temperature.

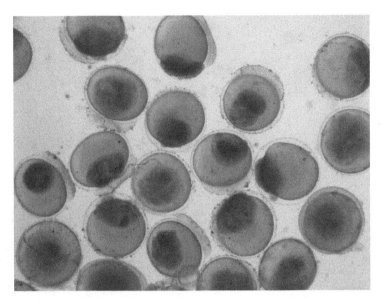

Fig. 11.12. Developing eggs

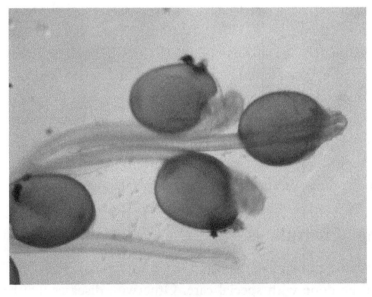

Fig. 11.13. Hatchlings with yolk sac

Rearing of catfish seeds can be divided into 2 parts. The hatchlings will take 15 – 20 days to attain 2-3 cm body length and 30-45 days to become fingerlings. The fish farmers can sell the hatchlings at both stages. The hatchlings need not to be fed for 3-4 days, because it will take nutrients from the yolk sac. After 3-4 days, it can be fed with live feed organisms. Initially it should be fed with hatched Artemia, then at later stage it can be fed with daphnia, moina, and tubifix worms. At the same time it can be slowly replaced by feeding with pellet feeds.

The catfishes hatchlings can be reared in cement tanks with special care. If the tank is constructed with 10m^2 then 20-25 cm water depth is required. For 1m^2 area, 5000 – 10,000 hatchlings can be stocked. After 1 week of rearing, it should be reduced to 1/3 part. At the initial days it should be fed with Artemia. Also it can be fed with boiled hen's egg yolk and daphnia. After growing for 4 days, it can be fed with tubifex worms. Almost special care should be needed for feeding the catfishes. Catfish hatchlings can also be cultured in ponds, swamps, etc, by constructing hapa or cages, according to our interest. 2000 – 3000 hatchlings can be stocked in 1 m^2 area.

Best feeds for rearing of catfish hatchlings

1. *Artemia nauplii:* The hatched out *Artemia nauplii* forms the best feed for rearing of catfish hatchlings. It can be used as feed for 7-10 days. The Artemia cysts can be purchased from the market. Then it can be hatched out using sea water. If seawater is not available, it can be hatched out using the salt and water solution mixture (28g salt in 1 litre water). In 1 litre of water, 3 g Artemia cysts can be hatched in the bucket, provided with 40 -60 watt bulb. It should be oxygenated with aerator. After 24 hrs, *Artemia nauplii* will be hatched out from the eggs. If the naupli has to be collected, then aeration should be stopped,so that unhatched eggs and eggs shell will float on the surface. Then it can be easily

drained using filter and *Artemia nauplii* can be collected. About 2 lakhs *Artemia nauplii* can be obtained from 1 g Artemia cysts. For growing 1 lakh catfish seeds, 25-30g Artemia cysts is required per day.

2. **Tubifex worms:** After 7 days of rearing, tubifex worms can be used as feed for catfish hatchlings. At first the tubifex worms should be washed in running water, then it should be cut into small pieces for feeding the catfish hatchlings. If it is not cleaned and used, then bacteria and other microorganism from tubifex worms will cause diseases in catfish hatchlings.

3. **Hen's egg:** The hen's egg yellow yolk and white yolk can be taken in equal quantity and mixed with milk powder. In addition to this, 0.5g vitamins and minerals can also be added. After mixing all the contents, it should be boiled in cooker. The boiled contents should be dried in cool place and preserved in refrigerator. When it is needed, it can be filtered and used as feed for hatchlings. The uneaten feeds should be removed immediately; otherwise it will pollute the tank and cause diseases.

Rearing of Catfish seeds

After rearing for 2 weeks, the grown hatchling (1000 Nos.) should be stocked in 4x1.0x0.5 m tank. Now, *Artemia nauplii* quantity should be reduced and mussels meat, mosquito worms, fish flour, pellet feed, tubifex worms, etc. can be increased in amount and used as feed. About 5g - 10g of these can be used as feed for 10,000 hatchlings. Now aeration can be reduced. Because after 10 days old, the catfish hatchlings can breathe oxygen from air. But slight water flow is always required and water quality also should be maintained and checked periodically. Within 10 -12 days the fishes will grow upto 2-2.5cm length and 0.8-10g weight. This is the suitable period to stock the hatchlings in the nursery pond.

It can be stocked in 50 m² area nursery pond. The pond depth should be maintained at 25-30cm. Pond should be fenced, to avoid the mixing of weed fishes. If it is a tank, then sand should be laid for 5-8 cm depth. The organic manure, cowdung can be used as 10 ton per hectare. Also, the inorganic manure, super phosphate can also be used as 20 g /m². Also the ¼ area of the pond can be slightly filled with *Eicchornia*(aquatic plant). It can be used as feed and also it will give shade in the pond. In nursery ponds the catfish hatchlings should be stocked at the rate of 100-500 numbers/m² and grown for 2-3 weeks. In nursery pond, the fishes will grow about 4-5 cm length and 10-20g weight and also it will get 45-70% survival rate. After growing in nursery ponds for particular period, it should be completely harvested and transferred to the grow-out ponds and stocked.

12

BROODSTOCK DEVELOPMENT, SEED PRODUCTION AND HATCHERY MANAGEMENT OF TILAPIA

Tilapia farming is getting more popular throughout the world. Recently, Tilapia is being cultured both in freshwater and brackish water regions in Asian continent. It is also cultured in the farm of extensive, semi extensive, and intensive systems. In world fish production, Tilapia culture stands next to carp culture. The main reason of this contribution was due to GIFT (Genetically Improved Farmed Tilapia) which was developed by World Fish Centre. Globally, Tilapia is being cultured in about 140 countries. Among the developing countries in Bangladesh, there are more than 400 tilapia hatcheries. From these hatcheries, they are producing about 40 million all male tilapia seeds using hormone treatment every year.

The success of any tilapia farming operation depends heavily on the quality of seed stock used. The main concern of farmers is to procure quality seed stock that can grow to marketable size within the shortest possible time. Farmers can procure seed stock from reliable reputable hatcheries, or operate a hatchery themselves to be assured of good quality seed. Another option is to obtain all-male stocks or genetically improved stocks from licensed farms, agencies, or research institutions.

Tilapias are normally stocked in grow-out enclosures at sizes ranging from 37 mm to 46 mm 'fingerlings'. Hatchery operators thus have to nurse 'fry' to the marketable fingerlings.

Operating a tilapia hatchery requires technical expertise. Knowledge in proper brood stock management is necessary. Genetically inferior or poorly managed breeders may produce poor quality fingerlings that grow slowly, have deformities, are diseased, or mature too early. On the other hand, good quality fingerlings have the following traits:

- Fast growth in terms of length and weight
- Robust body
- Normal colour
- No deformities
- Efficient feed conversion

Information regarding other criteria such as sexual maturity, resistance to diseases, and social behaviour may be obtained from the hatchery operator. It is also important to obtain pertinent information on the seed stock number, age, strain, ancestry or parental cross from which they were derived, and survival rate from hatching to nursery stage.

Brood fry requirement

It is suggested for those who want to start a large hatchery should start with new and the best original stocks. Normally hatchery operators who want to establish large hatcheries and expand further, they should start from 20,000 – 30,000 fry as a founder stock and keep them continuously in well-managed tanks or hapas in ponds. Total number of brookstock requirement can also be calculated based on the production target assuming an average 500 eggs can be collected from a female per month, and eggs have about 30% overall survival until being ready to be sold. If a hatchery has a target of producing/supplying 1-2 million fry per month, the number of mature brood fish required would be:

No. of mature females required $\quad$ = 1,000,000/ (500 x 0.30)

$\qquad\qquad\qquad\qquad\qquad\qquad$ = 6,667 females

No. of mature males (1:1 sex ratio) = 6,667 males

Total number of mature brood fish = 6,667 Females + 6,667 Males = 13,333

Assuming 75% survival during nursing, 90% during second nursing, 90% during maturation and 95% during handling, the number of brood fish fry required would be:

No. of broodstock fry required = 13,333/ (0.75 x 0.9 x 0.90 x 0.95) =23,102

The final number of broodstock fry to be obtained would be 25,000 considering the transport and handling losses.

Broodstock procurement

The required 25000 swimming up fry can be packed in 7 cartons / foam boxes. Each carton contains 4 plastic bags and each bag contains approximately 900 swim –up fry (size 0.02 g or bigger) in about 2 L of water and 3 L of oxygen. If the procurement is from abroad (cross – border movement), a health certificate of the brood fry based on the samples tested within a week of departure is needed. Normally, certificate is issued by the government agency in the exporting country. In addition, a certificate of origin may also be necessary. Some countries do not allow importation of exotic live species while others may issue licenses to importing parties with the provision of strict rules and regulations including a long process of quarantine and surveillance. Well-conditioned fry before packing i.e 24 hrs or more without feeding maintained in clear water can survive up to 36 hrs of travel (land and air). Therefore, brood fry can be imported from anywhere in the world.

Facility preparation and releasing fry

A small nursing hapa of $10m^2$ with a net protection against birds or a concrete tank of about $5m^2$ area outdoor or indoor is adequate for the arrival of new brood fry. The hapa and tanks have to be prepared (drained /cleaned / disinfected /limed) well in advance so that they shouldn't have any other fish or living organisms which may potentially harm or predated the new and young fish. If tap water is used to fill up the tanks, the water must be filled at least 3 days before to ensure that there is no chlorine remained in the water as the chlorine may kill the young broodstock. If present,

- Open all cartons and take all plastic bags out without opening the bags. Place all plastic bags to float on the water surface for about 15–20 min so that the water temperatures inside the bags will be similar to the temperature of the pond/tank water.

- Open the bags and gradually add the pond/tank water into the bags before finally releasing all the fry into the water. This is to avoid drastic change/shock to the swim-up fry due to differences in water temperature and other parameters.

- Broadcast some rock salts into the water inside the hapa as stress releaser and start feeding from the following day only.

Brood fish fry rearing and maturation

As broodstock fish are the key element of the hatchery, their quality and management affect the quality and quantity or seed production, which are ultimately the major indictors of success of failure of the hatchery. Feed the fry with fishmeal powder (or mixing rice bran up to half) to a satisfaction level (or 20 – 30% body weight/day for the first month and then reduce to 15% second month, and then 10% third month or after) separated into 4-6 times a day or every 3-5 hours. Table 12.1 gives a summary of management methods.

Table 12.1: Tilapia stocking, feeds and feeding

Month	Stage	Stocking	System	Feeds and feeding
0-2	1st nursing	1000	Hapa	Rice bran & fish meal (3:1),5 times daily to satiation
2-4	2nd nursing	200	Hapa	Rice bran & fish meal (3:1),5 times daily to satiation
4-6	Maturation	1	Pond	As above or pellets (30%CP*), twice daily to satiation

- If the pond has bird protection cover, and pond has no other fish, the hapa can be removed and released the fry in the pond water after about 3-4 weeks. But if not, they can be still kept in a larger hapa (20m^2) or split into two hapas of 10m^2 for another month. Again they can be separated gradually reducing the density by half each month.

- Fish will grow fast in the pond, if the water is green created by fertilization at the rate of 60kg Urea and 62kg TSP/DAP per ha per week.

- When they are around 100g size, which may take around 6 months, they are ready to stock into the breeding hapas (normally 60m^2 size) with 200 females and 160 males per hapa.

- Eggs should be harvested every week and incubated artificially.

Individual selection

- Tilapia can be used for breeding at the age of 6 months or when they are about 100g.

- Under or oversized fish should be discarded for breeding.

- Fish with good shape and attractive color need to be selected. Males with a pink colour on their dorsal and caudal fins are preferably chosen.

- Broodfish can be continuously used until the age of 2-3 years, depending on their size. Fish bigger than 300g are generally difficult to handle during seed harvesting. Therefore, they are isolated and transferred to fattening pond / tank or hapas so that they can be sold in a month which generates an additional return.

Sexing and Counting

- Separation of males and females are done by looking at the genital papillae or the vent. Females have red and round vents with short papillae (Fig 12.1), while males have long and pointed papillae (Fig 12.2). Males have only one pore in the papilla from where urine and sperms are released while females have two pores; one for releasing ova /eggs and another for urination

- Males and females are counted and kept separate for about 2-3 days so that they can be transferred as required to stock into the breeding ponds, hapas /cages or tanks.

Breeding or Spawning

- Male and female are stocked at 1:1 ratio. Stocking density varies with facilities. Normal stocking density of broodfish is 2 fish m^{-2} (3:3) in hapas and 10 fish m^{-2} (5:5) in tanks.

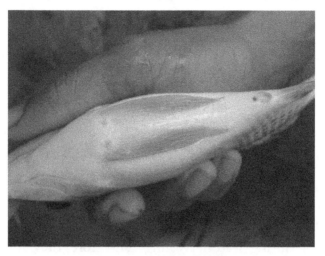

Fig. 12.1. Tilapia female

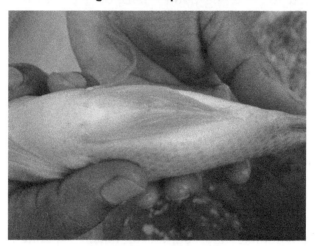

Fig. 12.2. Tilapia male

- As tilapia females spawn once in about a month, they need to be adequately supplied with nutrients throughout the breeding period. Floating pellets (Fig 12.3) are recommended for broodfish if available and affordable. Resource poor farmers can use simple rice bran alone or in combination with oil cakes and / or fish meal. The type of feed may depend on the breeding system used. In green water, broodfish are fed on a diet containing 25% crude protein at

0.6 -0.8% biomass per day split into two meals. On the other hand, in clear water system, 35% CP diet is used to feed broodfish at 2% biomass twice daily.

Fig. 12.3. Tilapia pellet Feed

• There are two systems of egg incubation, artificial and natural, which are described below

Broodstock replacement

Broodfish should be maintained in a controlled environment (tanks or hapas in ponds) so that they will not crossbreed with wild fish. Purity of the stock is very important. As large broodfish are difficult to handle during egg collection, they have to be replaced frequently; normally every 2-3 years or when they get bigger than 350g. They should be completely replaced following the all-in-all-out system. When replacing old stock, pond should be dried and hapas should be cleaned. New broodfish should be stocked in clean hapas installed in well-prepared ponds.

After procuring brood fry from outside or selecting from the on-farm hatchery broodstock fry are nursed normally. However, special care should be taken to avoid contamination from feral tilapias and to ensure that they have sufficient nutrients required for the somatic and gonadal growth. If broodstock are reared in clear water, a higher protein diet (35% crude protein) is required. But for the green water system a 25% crude protein diet would be sufficient.

Seed collection and incubation

Fry and egg collection

Fry collection

- Tilapias show a high degree of parental care to their eggs and fry. Females of the mouth brooding tilapia incubate eggs in their mouths (Fig 12.4) until the young can swim independently (swim-ups). These free swimming fry can be collected from the edges of the pond, hapa or tanks at an interval of 7-21 days using long scoop nets.

- This system is cheap and easy, but net fry production per unit area of space is very low because it is not possible to collect all the fry from the system. Therefore, it is often called the "partial harvesting method". Survival is also low due to predation and adverse environmental conditions. Furthermore, fry vary in size and age; therefore, if sex-reversal technique is to follow, partial harvesting does not yield good results.

- If harvesting is done more frequently e.g.2-3 times a day, a more uniform age and size of the fry could be obtained.

- Grading of fry is therefore important in partial harvesting which can be done using various sizes of mesh.

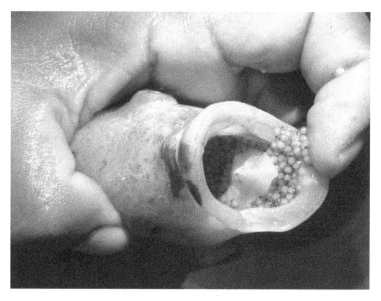

Fig.12.4 Collection of eggs

Fig. 12.5. Treatment of eggs

Eggs and yolk-sac larvae collection

- For a large scale hatchery, fertilized eggs (Fig 12.6) or yolk-sac larvae should be collected from the mouths of brooding females once every 5-7 days.

- The fish should be gathered to a corner of the breeding hapa. Then two hand nets (large mesh and small) can be used in conjunction for scooping up the broodfish. Eggs can be dislodged by putting a fore finger in the mouth of brooding females and shaking to release eggs or yolk-sac fry which are collected in the small mesh net. They are then transferred to plastic bowls with sufficient water to remain submerged.

- During harvesting, eggs or yolk-sac larvae are separated by stage (these are arbitrary stages based on the developmental extent which is observed, Stage I – just fertilized yellow in color with no spots, Stage II – with eye spots, Stage III – darker in color and with small tail and protruding eyes, Stage IV – with longer tail and head, Stage V –swim up fry)

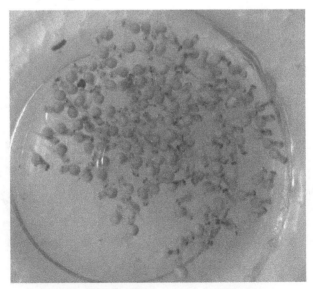

Fig. 12.6 Fertilized eggs of Tilapia

Eggs incubation and larval rearing

- The harvested eggs and yolk-sac larvae are transferred to the hatchery and washed with clean tap water (Fig 12.8) passing through a fine net (500 micron), disinfected with 40% formalin (4mL/2L = 200ppm) for 1-3 min and acriflavin (1g/200L water) solutions followed by rinsing in clear water each treatment (Fig 12.5), weighed and then placed into down – welling incubator jars. The jars are supplied with slow sand filtered water with a constant flow (Fig 12.7).

Fig. 12.7. Tilapia hatchery jars

- o The flow rate is adjusted in such a way that all the eggs are gently churned or agitated constantly throughout the day and night.

- o Normally, freshly laid eggs (yellow eggs) take about 4 days to hatch. Each subsequent stage takes one day less to hatch.

- o Immediately hatching takes place, yolk-sac fry are transferred to trays supplied with re-circulated clear water.

Fig. 12.8. Egg incubation system

Fig. 12.9. Tilapia nursery rearing

o If the stage IV and V swim-up fry are harvested, they are directly transferred to tray system for larval rearing for about 7 days.

o Swimming fry are then transferred to the sex-reversal system (hapa-in-pond or tanks). Until then, fry are not fed. By the time, they are transferred and MT feeding starts, fry are 10-11 day old post-hatching. It is essential to start MT feeding from this age as the sex determination takes place during 11-15 days. When they start taking MT feed, the level of hormone in their body is raised which directs the sex towards males. If MT feeding begins late, or when they are older than 15 days, sex reversal will not be possible. This is one of the fundamental principles; the hatchery operators should keep in mind.

Fig. 12.10. Tilapia hatchlings

All male Tilapia production

Sex – reversal

Although there are a number of methods attempted to produce all – male tilapia fry, hormonal sex reversal technique has been the most reliable and common methods. The period of sex-reversal can be divided into two phases; 21-day hormone treatment phase followed by 10-30 day nursing. Feeding young fry (Fig 12.11) with a male hormone (17α – methyltestosterone) produces phenotypically all – male fry (not genetically). It stops breeding in grow –out systems and males grow about 25% faster than females. For sex-reversal, 60mg of 17α- methyl – testosterone (MT) is mixed with 1kg of fish meal together with 20g of vitamin C or vitamin mixture.

Fig. 12.11. Hormone mixed feed for Tilapia sex reversal

Stock solution

As the hormone is insoluble in water, ethyl – alcohol (organic solvent) is necessary to dissolve it. The alcohol then helps spread the hormone evenly in each particle of feed / fish meal then it can be evaporated off easily and quickly under room temperature. While making stock solution, normally 5g of MT hormone is dissolved in 1 L of ethyl – alcohol using magnetic stirrer. The volume is then made up 10 L (by adding 9 L) which serves as stock solution and can be stored for about 6 months in a refrigerator at about 7°C. The stock solution contains 0.5mg of MT hormone per litre of alcohol that means, 120 ml of stock solution is required per kg of feed /fish meal to raise the required dose of 60mg/ kg feed.

Preparing MT Feed

While preparing MT feed, 10kg of feed (high quality fish meal or shrimp starter diet) is churned in a mixer. Gradual pouring of 600ml of stock solution and then another 600ml of fresh alcohol is done. This process is repeated again doubling the stock solution and hormone. The alcohol is then evaporated by spreading the mixed feed under shade for about an hour. Feed should not be dried under intense sunlight because the hormone will degenerate. After drying, the feed should be packed in a plastic bag or kept in a container with a tight lid and stored in a room at a low temperature.

Feeding MT feed

Fry stocked at a density of 5600 m^2 in a hapa (3 × 1.8m^2) are fed with the MT mixed feed at 14, 30,50 and 84g each day for the period of Day 1-5, Day 6-10, Day 11 – 15 and Day 16 -21, respectively. Feed is divided into five equal portions and the fry are fed 4 times a day. Size of the hapas during the 21 days of hormone treatment for sex-reversal depends on the scale of production. The smallest hapa can be 5m^2, medium 10m^2 and largest 40m^2.

Table 12.2. To prepare hormone mixed feed for the sex reversal technique, hormone concentration for particular days is given below.

Sl.No.	Weight of hatchlings / Nos.	1-5 days	6-10 days	7-15 days	16-21 days
1.	100g / 10,000 Nos.	25g	50g	85g	140g
2.	200g / 20,000 Nos.	50g	100g	170g	280g
3.	300g/ 30,000 Nos.	75g	150g	250g	420g

Table 12.3. Hapa size and tilapia hatchling numbers to be stocked

Hapa size	5.4 m^2	10 m^2	20 m^2
Tilapia Hatchling Numbers	30,000	60,000	1,20,000

Grading and nursing

Grading of fry is very important after the initial development stages to minimize mortality caused by cannibalism and social dominance, as tilapia fry are aggressive in nature. The first grading is done one month after hatching, followed by second grading event a week later (or a day before sale). Fry are grouped into 3-4 categories such as small, medium, big and very big, depending upon the size. During nursing (10-30 days), fry should be kept at a density of 1000 – 2000 fry m^2 in the hapa (Fig 12.9). The size of hapa can be between 20-120m^2 depending upon the scale of production. Fry are then fed on a mixture of rice bran and fish meal (2:1) 4-5 times a day at about 25-50g m^2 per day.

Nursing

Fry smaller than 1 inch is susceptible to predation by carnivorous fish and birds,smaller fish are also less tolerant to poor water quality, which is a common feature of ponds with organic manures. Factors such as these increase mortality rates, particularly in the first month of growth. Therefore, if they are nursed for at least a month before stocking them into grow –out ponds, their survival can be increased considerably. Following are the steps for normal fry nursing method:

- Small individual ponds, or hapas in a pond, reservoir or lake, are ideal for nursing. The system should be well protected from birds and other predatory fish or animals.
- Small ponds (50-500m^2) should be well prepared, and should be drained and dried completely at least for a week. Ponds should be limed at a rate of 500 – 600kg ha^{-1} and filled with predator free water.

- A week before fry stocking, ponds should be fertilized with urea and TSP to develop a green colour. Afterwards uneaten feed and their excreta will maintain the green colour. Overly green water is also undesirable during nursing. Therefore, organic manures are not recommended.

- Stocking density during fry nursing is 500-1000 fry/m² in hapa or cage and 100-200 fry/m² in pond.

- Fry are fed with floating pellets or with a mixture of rice bran and fish meal (2:1), 2-4 times a day at about 5-10% of biomass or to satiation.

- Nursing period is normally about one month. During this period fry attain at least 1g size which is a suitable size to stock into the grow-out pond.

- It is necessary to manage the timing of fry purchase with the period of nursing, fry harvest, preparation of the grow –out pond and fry stocking. It is recommended that new fry should be obtained one to two weeks to prepare (draining, drying, liming, filling and fertilizing) a pond, during which fry can be nursed in other smaller ponds or hapas.

Advanced Nursing

The level of seed demand is unpredictable for different seasons. It also differs year to year for the same season. Tilapia culture is mostly dependent on rainfall, which is the main source of water to fill the ponds. It is common to have large numbers of unsold fry during the dry season. Similarly, fry can also be kept at high densities during the cold season, and sold early in the next warm season at higher prices.

Why advanced nursing?

- to hold fry during low fry demand
- to keep stock of fry for the high demand season

- to minimize feed and feeding cost

- to grow fry to a bigger size, this means more protection from predators

- to have high potential growth (compensatory growth)

- to fetch higher prices from bigger fry

Advance nursing is sometimes called as stunting and over-winter depending upon the purpose and the facilities used.

Stunting

Intentional storing of fish at high densities with minimum feeding to keep fish alive is stunting. This can be done at any time of the year. Hapas or cages in ponds are used for stunting. About 2000 fry m^{-2} can be stocked in one hapa or cage and fed with a mixture of fish meal and rice bran (1:2). Feeding rate is maintained at 3, 2 and 1% biomass day $^{-1}$ for the first, second and third month, respectively.

Fry need to be prepared for transport before they are packed. The main problems during fish transportation are shortage of oxygen, high ammonia production from excreta, high temperatures (in hot seasons or areas), and the stress due to handling. In order to avoid these problems, the following steps are helpful:

- Fry harvest should be done at least one day before sale

- Fry are harvested from the hapas /cages or ponds using scoop nets

- Fry /fingerlings (Fig 12.10) are graded and kept in separate hapas based on their size.

- After grading, fry are counted and transferred to containers /tanks filled with the water from conditioning tanks. They should be transported to conditioning tanks immediately.

- Fry kept at high density in hapas in packing tanks supplied with clear water using sprinklers alone or with air-stones.

- Before putting fry into the conditioning tanks, they are dipped in a formalin bath (400 mL/in 100 L water) for 5-10 min to make sure parasites or other disease causing organisms are not transferred to the conditioning tanks and to the farmers field.

- It is essential to condition the fry /fingerlings by starving overnight so that they empty their stomach and intestines. This reduces the production of excreta in plastic bags / transporting containers. This avoids mortality during long transportation periods.

Genetically male tilapia (GMT) or YY technology

- Direct hormonal masculinization might not appear to be a viable technique in tilapia and might have adverse environmental impacts or consumer reaction in near future, in that case the indirect method of producing monosex all males (i.e. YY males) by combining both sex-reversal and/or genetic manipulation of the sex determining system will once be the alternative choice of the commercial seed producers.

- This technology was developed in the late 1980s by the University of Wales, Swansea, Wales, UK in collaboration with the Freshwater Aquaculture Center of the Central Luzon State University in Munoz, Nueva Ecija, Philippines. The collaborative work was implemented to produce all-male Nile tilapia populations on a large scale. All-male populations are obtained through a series of genetic manipulation, feminization, and progeny testing to produce novel YY males. The genetically male YY genotypes, when mated with normal females XX can produce all-male XY progenies. This method has a mean success rate of 96.5% males.

Acclimatization to salinity

- In order to grow tilapia in brackish or saline water, fry / fingerlings should be acclimatized.

- For acclimation of fry to brackish water, salinity should be increased by 2.5 ppt daily until reaching the same salinity level of the grow –out farm

- Red tilapia fry acclimate to brackish water more quickly. Nile and blue tilapia fry need gradual acclimation to brackish water before transferring to these systems.

Quality assessment and certification

- Fry quality can be assessed by observing the movement, color, shape, size, responses to feed and strangers.

- Fry quality can also be tested using the salinity challenge test (select 100 fry, submerge in 24ppt salt solution for two hours; repeat it twice or thrice; observe survival, it should be about 50%, the batches with higher survival rates can be considered as better quality which increases market value).

- Fry can be carried in earth / metal container (e.g., Bangladesh) but polythene bags (5L) are easier and popular. They are filled with 2 litre clear water with 3L oxygen.

- Normally 1000 fry are packed per bag. As counting of all the fry is time consuming and tedious job, three batches of 1000 fry are counted and weighed. The average weight is used to estimate the remaining fry and transferred to each bag. Another way of estimating number is by volume; a small cup can be used to record the samples of 1000 fry and the same volume can be used for the rest of the packing process.

- Number of fry should be reduced to 500 if they are transported for longer over 12 hrs.

- After the 1000 fry are isolated they are transferred into the polythene bags. Excess air should be removed and refilled with oxygen from a cylinder before being tied and sealed with strong rubber bands.

- Water and oxygen can be changed during transportation if the distance is over 12 hrs, based on requirement and available facilities.

- About 10% of fry can be lost during conditioning and packing. Sometimes high mortality rates can be experienced in cases of electricity failure or any other causes that create adverse conditions such as low temperature, low dissolved oxygen and physical stress.

- Well – conditioned fry packed with oxygen can be transported to any part of the world, and for upto 18 hours by road.

- Fry transport should be avoided during day time especially in hot and dry seasons. If necessary, plastic bags should be covered with thick sacks. Water can also be spread over the bags to keep temperatures down before and during transportation.

- Any means of transport can be used to carry fry. A simple bicycle is useful to carry fry for short distances as is done in Bangladesh and Vietnam. Vehicles like cars, pick – ups and trucks can be used to transport fry for a full day by road.

- Fry can be sent to anywhere in the world by air cargo at low temperature (up to 36 hrs including local travel). Usually 4 fry bags are packed in a foam box. The foam box should be labeled with name and proper address of the receiver.

- In case of cross boarder transportation, a health certificate issued by the government agency to guarantee that fry are free of parasites and disease causing organisms is required. At the same time, importing farmers may need license and to follow quarantine regulation.

- Fry of fingerlings can also be transported using plastic or steel tanks with an aeration system.

- In Bangladesh farmers use aluminum or earthen vessels carried by bicycle or manually. They physically agitate the water during transportation to introduce oxygen from the atmosphere.

- When releasing fry into the culture system, the plastic bags should first be allowed to float in the receiving tank/pond so the fry gradually acclimate to the new water temperature. In cases of transportation in tanks or other vessels, water from the pond can be added to the container, gradually replacing the original water and decreasing the gradient which may affect the fry.

13

WATER QUALITY MANAGEMENT IN AQUACULTURE

For any successful aquaculture activity, water quality is the most important factor and some of the parameters like, salinity, pH, hardness, etc. should be initially checked before filling. Temperature, dissolved oxygen, ammonia and nitrite are the most important factors.

To obtain favourable economic returns and rapid fish growth, fish are stocked at relatively high densities and fed a high quality diet. The metabolic wastes accumulating from the intensive feeding of the fish represents considerable nutrient enrichment of the water. The wastes added to the pond stimulate the growth of aquatic plants. Usually phytoplankton and the combination of high fish standing crops and dense plankton communities exert a tremendous oxygen demand. The further intensification of catfish pond culture is largely limited by the amount of feed that can be added to the pond without causing economically unmanageable oxygen depletions. The goal of water quality management is to regulate the physical, chemical and biological environment of the culture system so that conditions favour the optimum growth of the fish crop.

1. Water quality and fish feeds

Changes in water quality occurring during fish culture directly or indirectly are the result of the feeds added to the pond to increase fish growth. To understand why feeding the fish pollute pond waters, it is necessary to examine the fate of the nutrients contained in the feeds. Only about 30% of the nitrogen and phosphorus originally contained in the feed is removed from the pond when fish are harvested. The rest is lost to the water as excretory products. Obviously, the quantity of nutrients reaching the water increases when feed conversion is poorer. For example, with a feed conversion ratio of 2.0 rather than 1.3 the amounts of nitrogen and phosphorus lost to the water more than double. Cultural practices resulting in more efficient use of feed by fish may also result in fewer water quality problems.

Almost all the nitrogen and phosphorus input was from the feed applied. Most of the phosphorus ultimately lost was adsorbed by the pond muds. The processes of adsorption by muds (for phosphorus) and denitrification of ammonia volatilization (for nitrogen) serve to moderate their concentration in the water column. Considerable quantities of these substances are also assimilated by phytoplanktons. At any one time, between 50 and 90% of the total nitrogen in the pond water and 75 – 95% of the total phosphorus is in the particulate fraction, presumably most in phytoplankton cells. The rapid uptake by plankton organisms of nutrients in fish excretory products is particularly important in the case of nitrogen where ammonia is potentially toxic to fish and the uptake and use of the ammonium ion as a nutrient source by phytoplankton generally serves to maintain the concentration of total ammonia-nitrogen at moderate concentrations in the pond water.

2. Water quality and feeding rates

As the standing crop of fish increases during the growing season as a result of fish growth, the average daily feed allotment per unit pond surface area increases. Thus the amount of substances reaching the pond water in fish excretory products also increases with time. This results in a characteristic temporal change in certain water quality variables. Concentrations of chlorophyll a, Chemical Oxygen Demand (COD), Biological Oxygen Demand (BOD), organic nitrogen, total phosphorus, nitrate, total ammonia, and filterable orthophosphate gradually increased during the growing season as feeding rates increased in catfish culture ponds. Abundance of phytoplankton (indicated by chlorophyll a concentration) increased in response to the increasing amounts of nutrients available for plant growth during the warm summer months. As phytoplankton abundance increased, the quantities of organic matter (COD), organic nitrogen, and total phosphorus increased. Concentrations of total ammonia increased because ammonia is the principal nitrogenous waste product of catfish and ammonia is continuously excreted by fish at rates roughly proportional to feeding rate. As the ammonia was nitrified, concentrations of nitrate increased. The accumulation of nutrients and organic matter and the problems with low dissolved oxygen and toxic nitrogenous substances may be more frequent and severe in ponds that are not drained annually than in those that are emptied and refilled each year.

Some water quality parameters are largely unaffected by the addition of the fish feeds. Concentration of total alkalinity, hardness, and chloride are initially dictated by the quality of the water supplying the pond. Changes in the concentrations of these variables are largely the result of the physical processes of dilution by rainfall or concentration by evaporation. Parameters such as these are termed 'conservational' in that they are not significantly affected on the gross level by biological activity. Although changes in conservative parameters are relatively slow, they can be of considerable practical importance. For example, the concentration of chloride (Cl) in the pond water influences the toxicity of nitrite (NO_2) in catfish.

3. Water temperature

The most important physical parameter is water temperature. In tropical countries, fishes grow well at 25°C– 32°C. Also in summer seasons, it can grow well in some ponds at 35°C. In this higher temperature also, some fishes like tilapia and catfishes grow well. But tilapia will stop breeding at this temperature. We have to be careful in maintaining the temperature because the fish body temperature will change according to the water temperature.

4. Turbidity of water

It is also one of the important parameter of water. Generally the pond water turbidity is determined based on the quantity of planktons present in water. The pond depth can be measured using secchi disc (Fig.13.1). The light penetration should not exceed 33 cm in the pond.

Fig. 13.1. Secchi disc

5. Oxygen

Oxygen consumption rates by fish vary with dissolved oxygen concentration, feeding status, weight of fish, and water temperature. Oxygen consumption becomes less as the dissolved oxygen concentration decreased. Well-fed fish consume more oxygen than by fasted fish. Oxygen consumption also increases with temperature increase. Lean fish will consume less oxygen than fat fish. Prolonged exposure to sub lethality low concentrations of dissolved oxygen is harmful. The feed consumption will be reduced often when dissolved oxygen concentrations are low. Food conversion values are higher (feed used less efficiently) for fish in ponds with chronically dissolved oxygen concentrations than in ponds with higher dissolved oxygen concentrations.

Diurnal changes in DO in an infertile pond are small but in a fertile or eutrophic pond may vary from zero at dawn to double super-saturation in mid afternoon. Wide daily fluctuations in dissolved oxygen concentrations typically occur in fish culture ponds. Fishes will not feed well where dissolved oxygen concentrations fall below 1 mg/l. Continued exposure to low dissolved oxygen may predispose fish to bacterial infection. Under pond culture conditions, carbon di-oxide and ammonia concentrations are often high where dissolved oxygen concentrations are low. The minimum concentration of dissolved oxygen tolerated by fish is obviously a function of exposure time. A fish might survive 0.5 mg/l dissolved oxygen for a short time but not for several days. Further, the minimum tolerable concentration of dissolved oxygen will vary with fish size, physiological condition, concentrations of solutes, and other factors.

Table 13.1. Dissolved oxygen requirements for warm water pond fish

Sl. No.	Dissolved oxygen	Effects
1.	< 1mg/l	Lethal if exposure lasts longer than a few hours
2.	1 – 5 mg/l	Fish survive, but reproduction poor and growth slow if exposure is continuous
3.	>5mg/l	Fish reproduces and grow normally

Generally in aquaculture, systems the presence of O_2 in the pond is more important. Because, if the fishes were stocked in high stocking density then O_2 depletion will occur. Amount of O_2 in water will depend on water hardness, temperature, quantity of planktons, etc. If O_2 is low in water, then it can be overcome by using aerators.

The fishes like murrels and catfishes have special breathing apparatus. So, it is easy for them to breathe the oxygen present in atmosphere. Also, these fishes frequently visit the surface of water for this reason. So these fishes can survive and withstand low amount of O_2 in pond. However, fertilized eggs need lot of oxygen for the development. If the fertilized eggs were kept in oxygen below 2-5mg/l of water, then mortality of eggs will occur, because developing eggs need enough oxygen for development.Healthy fishes can live in water with 2mg/l of oxygen. But it will show stunted growth. Also these fishes will be easily prone to diseases.

A pond with light plankton bloom normally has dissolved oxygen concentrations near saturation-ideal conditions for respiration. Metabolic wastes promote plankton growth; so as feeding rates increase, plankton blooms flourish and dissolved oxygen regimes decrease in stability. Fish are exposed to high dissolved oxygen concentrations during the day and low concentrations at night. The influence of these fluctuation on fish growth is poorly understood, but fish converted feed best when oxygen, concentrations are at or above saturation most of the day and do not fall below 25% if saturation at night.

Phytoplankton die-offs can cause severe depression of dissolved oxygen concentration. Die-offs are characterized by sudden death of all or a great portion of the phytoplankton followed by rapid decomposition of dead algae. Dissolved oxygen concentrations will decline drastically, and they may fall low enough to cause fish kills. The causes of phytoplankton die-offs have not been determined exactly, but in catfish ponds they usually involve dense surface, scums of blue-green algae and occur on bright days when dissolved oxygen concentrations are high, carbon-di-oxide concentrations are low, and pH is high, Emergency aeration will prevent a massive fish kill in the pond.

Aeration

Aerators provide oxygenation by increasing the area of contact between air and water and the amount of turbulences. Aerators increase the area of contact between air and water by (1) agitating surface water, (2) releasing air bubbles beneath the surface, or (3) Both. While oxygenating water, aerators also import energy to water and cause horizontal and vertical mixing. Diffused air aerators employ an air blower or air compressor and porous tubes to release air bubbles at the pond bottom. The efficiency of oxygen transfer is related to bubble size – small bubbles offer a greater air-water interface than large bubbles. Water depth also increases efficiency, because the deeper the bubbles are released, the greater the contact time between bubbles and water.

Surface aerators pump, spray or splash water with the air. Water is broken in to turbulent, thin layers and drops that have large surface – to- volume ratios. Vertical turbines, paddle wheels and pump – sprayers are common types of surface aerators. Propeller-aspirator-pump aerators release fine bubbles of air which are mixed in to the water by turbulence created by the propeller.

The Paddle wheel aerator is usually positioned so that paddles extend 20-30 cm into the water and the paddle wheels are rotated at about 120 rpm. They splash water into the air and cause strong water currents (Fig.13.3). Under ideal condition, the water coming into a fish pond or tank should be fully saturated with oxygen. The ability of water to take up oxygen depends on a number of factors, including temperature, pressure and dissolved salts. The higher the temperature, and/or salinity, the dissolved oxygen in the water is less. The oxygen consumption of fish per unit mass increases with increasing temperature, decreasing fish size, and increasing feed consumption. The magnitude of the standing crops of phytoplankton, fish and other pond organisms also affect dissolved oxygen and carbon-di-oxide budgets. Dissolved oxygen concentration fluctuate more diurnally and are lower at dawn as the growing season progress. Similarly, carbon di oxide concentration at dawn increases with time.

Fig. 13.2. Paddle wheel aerator

6. Carbon-di-oxide and pH

Diel fluctuations in carbon-di-oxide concentrations are essentially the opposite to those for dissolved oxygen. Carbon-di-oxide is produced during respiration and consumed during photosynthesis. Thus, there is usually a net loss during daylight and a net gain at night. Although carbon di oxide is not appreciably toxic to most fish, it antagonizes the uptake of oxygen. When dissolved oxygen levels are marginally low, fish may survive at low carbon-di- oxide concentrations but suffocate if carbon-di-oxide concentrations are high. The use of aeration is the most practical means of pretending the undesirable effects of high carbon-di-oxide concentrations.

pH

The extent of pH fluctuation in pond water depends on the amount of carbon-di oxide removed or added by the pond community and the total alkalinity of the water. pH will fluctuates greater in waters with abundant algae.

7. Water hardness

Good aquaculture ponds should have the water hardness of 6.5 to 9.0 ppm. Initially from 5-6.5 hardness fishes will show stunted growth. If the hardness increases above 9.0, then mortality is observed due to higher hardness in water. If the water hardness is found low, then heated lime should be applied.

8. Ammonia and Nitrite

Ammoniacal substances are produced as excretory products by fish, and also by breakdown of decaying foodstuffs in the pond. They are found in water in one or two forms: free ammonia (NH_3) and ionised ammonia (NH_4^+). The free ammonia is by far the more toxic of the two, and concentrations as low as 0.017 mg/l NH_3 are likely to be toxic to the gills and air breathing organ, resulting in reduced growth and poorer resistance to oxygen and other stresses. The relative proportions of free ammonia and ionised ammonia depend on the pH and temperature of the water, the higher the pH the greater the proportion of free ammonia (i.e. toxic ammonia). Ammonia in the gaseous state is only partially removed by agitation of the water and the concentration in the water is related to pH, temperature and water chemistry. Unfortunately, it is difficult to control the occurrence of their toxic substances in typical commercial culture ponds except by reducing the amount of feed added.

Generally in aquaculture ponds, if the fishes were grown in high stocking density, then aerator should be used. It is wrong to decide when there is aerator, more quantity of fish can be grown. Because the nitrogenous waste present in pond water, will affect the fish health and growth. In aquaculture ponds, ammonia will get developed due to two reasons.

- Fishes will digest the proteins in the food and release it as ammonia.
- The ammonia which get excreted from the fish body will get vary according to the amount of feed taken.

If the fish took 1 kg feed, then it releases 0.03 g ammonia nitrogen. Thus it will form ammonia as 100g/ha pond area. Hence, further when the fish grows ammonia released amount will also get increased. Also there is a chance (i.e) the NH_3 can enter the fish body and cause death. If 1mg of ammonia present in 1 litre of O_2, it will not cause any disease. If it is exceeding this level then mortality will occur.

Ammonia will get converted to nitrate by using bacteria. This nitrite level will be more in catfish ponds. It will cause major effects in fishes. The nitrite present in pond water will get enter into the fish blood circulation and cause disease in fishes. Also this nitrate will change the RBC and level and can cause diseases in fishes.

9. Toxic substances

In addition to ammonia, the decay of fish wastes and uneaten food can produce toxic substances. Hydrogen sulphide (H_2S) is produced at certain pH which is stressful, though not in itself particularly poisonous. Increased levels of CO_2, produced during the respiration cycle of phytoplankton, may also cause respiratory stress. In fishes levels of as low as 10 mg/ℓ have been shown to induce kidney disease. Suspended solid levels are always high in certain catfish ponds because of the stirring –up of mud by the fish, but the fish are usually resistant to damage from this source because of the mucus they produce on the gills. Excessive organic solids, on the other hand, may increase bacterial levels, and deplete oxygen. Pollutants such as chemical sprays and insecticides from rice fields are known to be particularly toxic to fish, and industrial effluents from factories, which may also find their way into pond water supplies, may also be toxic. Another significant form of environmental stress occurs during sudden changes in water quality, for example during water exchange, or during extremely varying photosynthesis/respiration cycle. If solid waste substances are present more in fish ponds, then bacterial production will get increased and oxygen depletion will occur. If the pesticides used in agriculture farms were used in fish ponds it will also increase the production of toxic substances. If the industrial water is released in aquaculture ponds, it will also increase the toxic level in fish ponds. Besides this, during water exchange, we have to be more cautious. Hence partial water exchange is the best method for fishes in cultured ponds, to maintain water quality.The ideal range of water quality parameter for aquaculture is given in Table 13.2.

10. Stress and general adaptive responses

All animals are able to withstand unfavorable conditions to a certain degree. While husbandry stresses may not necessarily be fatal in themselves, they may render the fish more vulnerable to infections. Stresses which are particularly dangerous to fishes are grading or transporting in very high density. Severe traumatic damage may be produced by their spines, by stabbing or scratching each other. Holding in small tanks in bright sunlight, can allow very high temperature to develop, leading to stress. Improper handling of fry during transfer from nursery to grow-out facilities can cause considerable stress and subsequent mortality.

Table 13.2. Ideal range of water quality parameters for fish culture

Sl.No	Water quality parameters	Ideal range	Maximum concentration
1	Water temperature	25-30°C	32°C
2	Salinity	0.5 – 4 ppt	20.1 – 11ppt
3	Oxygen content	5-6 mg/l	2 – 10 mg/l
4	Water PH	6 -9	5 – 10
5	Water hardness	0 mg/l	1-7 mg/l
6	CO_2	0 mg/l	0 – 20 mg/l
7	NH_3	0 mg/l	0 – 0.2 mg/l
8	Nitrite	0 mg/l	Depends on chlorine amount

11. Stocking rates and mortalities

One of the most common problems in fish culture is the overstocking of ponds. Stocking beyond the limit causes mortalities, often by disease, often indirectly by poor environmental conditions, until the stock falls below the limit of capacity. While it is true that the dead fish are not completely wasted, the cost of fry or fingerlings makes them very expensive feed for the survivors. During infections disease outbreak, the situation becomes worse coupled high stocking density. As a result, very large number of stock can be lost. If the pond is stocked at the correct level, however, mortalities are normally kept low; the risk of disease is generally lower. Excessive stocking densities tend to reduce growth rates, due to parasitism and other factors and therefore the production period may also be prolonged.

12. To control bad smell in fish ponds

Generally, bad smell is felt over the fish ponds, it is because of the presence of planktons and bacteria. If fishes were sold in fish markets with the smell, no one is interested to buy fishes. Also it will not be sold for good price. Hence it is required to clean the pond to reduce the quantity of bacteria and planktons. The harvested fishes should be kept for 2 hrs in running water to eliminate the bad smell. Thus, finally bad smell will get rid from the fishes.

13. Recirculatory Aquaculture System

Recirculation aquaculture (Fig 13.3) is essentially a life support system for fish. It is generally defined as intensive aquaculture in which the water is reconditioned as it circulates through the system and no more than 10% of the total water volume of the system is replaced daily. In order to compete economically and efficiently use the substantial capital investment in the recirculation system, the fish farmer needs to grow as

Fig. 13.3. Recirculatory system

much fish as possible in his system. The level of intensity in recirculation aquaculture is expressed as weight of fish per unit of water. The upper limit of a system based on the atmospheric oxygen appears to be about 0.5 lb/gal (60 g/l). In systems that use liquid oxygen to boost production this can go to 0.75 lb/gal (90 g/l) or even higher. This is the equivalent of a 10 -12 inch (25 - 30 cm) fish living happily in a gallon (4 l) jar. Obviously, to keep so much fish alive in so little water requires a well designed system. Usually, fish die if overcrowded because they either 1) suffocate (this will occur in hours) or 2) poison themselves with nitrogenous waste (this takes longer). A properly functioning recirculation system must aerate the water in some way, adding oxygen and conversely removing carbon dioxide as well as removing the ammonia that fish excrete as a byproduct of the catabolism of protein. Before these two processes can be conducted efficiently, solid waste (feces and uneaten food) must be removed from the system. So, three things must occur as the water is reconditioned: 1) removal of solid waste, 2) gas exchange, and 3) removal of ammonia. These last two may be conducted in either sequence or together depending on the system. These three fundamental processes cannot all be effectively executed in the fish tank so the water must be recirculated, or moved, through different modules by a pump. NFDB and Department of Fisheries, Govt. of India is supporting many entrepreneurs and state government to adopt the farming system for a sustainable water management.

14. Biofloc Technology

Biofloc technology makes it possible to minimize water exchange and water usage in aquaculture systems through maintaining adequate water quality within the culture unit, while producing low cost bioflocs rich in protein, which in turn can serve as a feed for aquatic organisms. Compared to conventional water treatment technologies used in aquaculture, biofloc technology provides a more economical alternative (decrease of water treatment expenses in the order of 30%), and additionally, a potential gain on feed expenses (the efficiency of protein utilization is twice as high in biofloc technology systems when compared to conventional ponds), making it a low-cost sustainable constituent to future aquaculture development. Conventional technologies to manage and remove nitrogen compounds are based on either earthen treatment systems, or a combination of solids removal and nitrification reactors. Biofloc technology, on the other hand, is robust, economical technique and easy in operation. One important aspect of the technology to consider is the high concentration of total suspended solids present in the pond water. Suitable aeration and mixing needs to be sustained in order to keep particles in suspension and intervention through either water exchange or drainage of sludge might be needed when suspended solids concentrations become too high. Although it is a critical aspect of biofloc technology, detailed knowledge about selection and placement of aerators is still lacking. Future research should address this issue and could also investigate new concepts, such as the integration of biofloc technology in raceways, which might prevent solids build up through its proper system configuration. Construction aspects for biofloc technology ponds merely deal with aeration. So improving and fine-tuning of the design of these ponds in terms of water mixing and sludge control is needed. The technology can be easy used for high value fish seed rearing to produce good quality fish seed.

14

FISH DISEASES TREATMENT AND ITS PREVENTIVE MEASURES

Water quality parameters, giving more feed (i.e.) above the required level than the fish required, more stocking density, etc. are some of the major reasons to cause diseases in fish culture ponds. Bacteria, fungi, parasites and some nutritional deficiency in feed are also the reasons to cause diseases in fishes. They are explained below in detailed manner.

1. Bacterial diseases

Among the bacterial diseases, bacteria septicemia and bacteria wound diseases are the most important. This bacteria septicemia disease will be caused by *Aeromonas hydrophila*. This disease occurs due to poor water quality, parasites attacking the fishes and oxygen depletion. The fishes which are affected from this diseases will show the symptoms like lying down in bottom of the pond horizontally, staying in pond surface in vertical position, it will not take feed due to the loss of appetite. Its fins will be in bloody reddish and eyes will get bulged. To get cure from this disease, fishes should be treated with antibiotics, it can be given through feed or through the water bath treatment. Any one antibiotics of Teramycin, Oxytetracycline or Chloramphenicol should be given @ 5g/100 kg feed for about 2-3 weeks. If the fishes are not taking feed then it can be given through the bath treatment.

2. Fungal diseases

The most important fish disease causing fungi is Saprolegnia. This will cause thread like growth formation in fish skin and scales. These types of disease attacked fishes do not swim properly. It should be treated with 0.1 ppm malachite green or 3 ppm salt mixed solution for 20-30 min.

3. Parasitic diseases

Parasites are organisms which spend part of their lives living on or at the expense of other animals. Parasites may be found in all tissues of the fishes, but they are particularly common on the skin and gills, because these external surfaces are easily invaded due to the flesh and blood. In the wild, although fishes have many parasites, they are usually found in small numbers due to the extensive area of distribution and rarely cause any serious injury. In the fish ponds, however, they can build up to serious levels very quickly and result can be very serious.

i)   **Trichodina complex:** Among the most important protozoan pathogens for fishes, the species of *Trichodina* (Fig 14.1) and *Trichodinella,* which make up the Trichodina complex are important ones. *Trichodina* spp., a small saucer (–) shaped protozoa found on the skin and gills of all sizes of fishes and they cause severe problem to the fry and young fish. Affected fish have a darkened appearance and the skin is often seem to be grey and flaking off. They are inactive and generally do not feed. They show signs of irritation and may "flash", that is swim on their side for a second or two and rubs against the bottom as if to try to scrape parasites off. Formalin at a rate of 25 ppm in the pond is an effective control for this parasite.

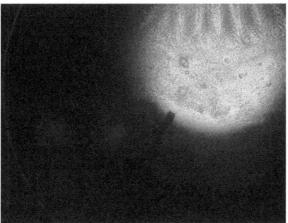

Fig. 14.1. Trichodina

ii) *Gyrodactylus* **spp:** These monogenean parasites have only been found on the skin but they are often present in significant numbers. They are viviparous. When the fish are stressed and there is skin damage, they can very rapidly build up in numbers. Gyrodactylids are up to 2 mm in length and can readily be distinguished from other monogeneans in skin smears under the microscope, by the absence of eye spots and the occurrence of the embryo in the mid-region of the body. Affected fish have dark patches over the body surface, with sloughing areas of skin. They generally do not feed, and the parasites can be readily seen in skin scrapings, often accompanied by trichodinids. Dipterex at a rate of 0.25 ppm or formalin at a rate of 50 ppm are very effective in controlling this parasite.

iii) **Dactylogyrids:** Dactylogyrids are very common on the gills of cultured fishes.They can multiply very rapidly and in such cases small numbers may overflow into the buccal cavity. At high levels they are likely to interfere with respiration and are often found in association with protozoan parasites of the gill. Infected fish may be darker and have poor appetite. The Dactylogyrus monogeneans are oviparous unlike Gyrodactylids. Formalin at a dosage of 25 ppm for 12-24 h can control this parasite.

iv) **Metacercariae:** These intermediate stages of the digeneans parasite life cycle are very numerous in tropical waters. They penetrate the skin or gills of affected fish and may also migrate through the body to localise in various other organs such as heart or peritoneal tissue. In farmed fishes, they are common on the gills. Heavy infections in farms may cause mortality in fry and possibly inhibition of organ function in adult fish.

v) **Myxosporidians:** Cysts of common myxosporidian, *Henneguya* sp. are found on the skin and gills of fishes in the form of small whitish nodules. Subsequently the intensity of infection becomes severe. When the cyst is squashed, large numbers of spores can be seen under the microscope. Individual cysts can be seen on fish of all sizes but large numbers generally occur on fry and cause serious problem. Occasionally very heavy infections have been reported resulting in heavy mortality. Other myxosporidian infestation include, *Myxozoma/Myxobolus* the cysts of which could be encountered in the gonads of older fish but only in small numbers. *Myxidium* sp. is found regularly in the bile ducts and gall bladder.

The life cycle is thought to be direct and infection is by ingestion of one of the myriad spores produced by rupture of a lesion of an infected fish. The static fish ponds with heavy mud/sediment are an ideal environment for this parasite, which tends to build up over long period of time. It is essential to remove dead fish to reduce the incidence of *Myxosporidia* but no satisfactory drug treatment is available.

vi) **Cestodes:** Tapeworms, either in the form of intermediate stages in the viscera, or as adults are occasionally found in fishes but not in significant numbers.

vii) **Nematodes:** Roundworms are found only in small numbers in fishes and are not known to cause any significant damage to the fish.

viii) **Argulosis:** It is caused by species of the genus Argulus, also commonly known as fish louse. Symptoms include small red patches on skin, gills and fins. Argulus parasite (Fig 14.2 and Fig. 14.3) has hook like

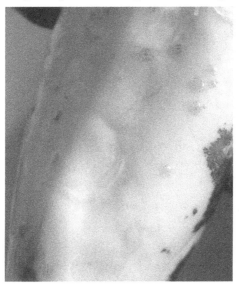

Fig. 14.2. Argulus

Fig. 14.3. Argulus Infection in koi Carp

limbs and a sucking feeding apparatus and can cause open wounds that can lead to secondary infection. Feeding sites are marked by haemorrhagic spots, hyperplasia of epidermis at the margins of wound. Sometimes, the mature parasite can be seen crawling on body surface or attached to the base of the fins or operculum.

ix) **Lernaeasis:** Lernaeasis is caused by species of the genus *Lernaea*. This parasite is commonly known as "anchor worm". The mature female parasite (5-15 mm long) penetrates the skin of fish and remains attached to the host with its head buried in the muscles and the elongated body protrudes outside with a pair of egg sacs attached to the host.

Different methods are used to control above parasitic infestation in fish

- Before stocking, bath fingerlings in 20 ppm Potassium permanganate ($KMnO_4$) for 15–30 min.
- Dip treatment with Sodium chloride (NaCl)/ Salt solution @ 5g/ l for 30 seconds.
- Apply Formalin @ 1.5-2.0 ml for 1000 lit water volume
- Apply Deltamethrin preparations (Trade Name: Buox (1.25%) or Ectodel (2.8%) 15- 20 ml/acre per meter water depth in three repeated doses at 5-7 days interval
- You can use Cypermethrin (Trade Name: Clinar EC:10) @15 ml/ acre per meter water depth.
- Ivermectin power (1% w/w) HITEK: mix 20-25g powder in 100kg feed, feed for 10 days.
- Emamectin Benzoate powder: mixed with feed @5mg/100kg biomass, feed for 10 days.
- Prepare the pond (before fertilization) with one round of parasiticide treatment (eg. Deltamethrin) before stocking of seed.

- Lift the seed from pond pre-treated with one dose of parasiticide (eg.Deltamethrin).

- Maintain optimum pond pH and total alkalinity (100 – 180 ppm) of water.

- Apply lime based on pH of water (7.5 – 8.5).

- Mix Ivermectin in feed for three days after three months of stocking.

- Deltamethrin treatment (three applications at weekly interval) 15-20 ml/acre per meter water depth in three repeated doses, after six months of culture or Lufeneuron @ 0.1 ppm in water for weekly once for 5 weeks.

- One more dose of Ivermectin after 9 months of culture, if required.

4) Nutrition deficiency diseases

i) **Vitamin C deficiency disease:** If Vitamin C is deficient in feed, this disease will occur in fishes. The affected fishes will have large skull bone. If the affected fishes have any wounds in their body then it takes long time to disappear. Also these fishes will have dull body and it cannot swim properly. To treat these kind of affected fishes is very difficult. It can be done by stop feeding the fishes for some days and water exchange should be done. After some days, Vitamin C is added in the feed can be given for complete curing of the fishes.

ii) **Open stomach disease:** If the fishes were stocked in high stocking density or more protein is added in the feed then this disease will occur. Especially, this disease will occur in catfish hatchlings. This disease affected fishes will have swollen stomach and swim very fast from one place to another. After some days of swelling, the stomach will get burst. This type of affected fishes should be stopped feeding for 2 days, after that the feed range can be slightly increased.

5. Preventive measures to cure fish diseases

Disease preventive measures are the most important thing in agriculture, Veterinary and fish culture. In aquaculture, the causes of fish diseases can be easily identified and it can be prevented earlier.

Recently, carps, catfishes and tilapias could be grown in large numbers in high stocking density. However, if it were grown beyond the limit, it will cause many problems in future. Initially it will affect the water quality (i.e), Ammonia (NH_3) and Hydrogen Sulfide (H_2S) will get increased in the water. The pH of the water will also get changed. Many changes will occur in gills and skin regions of fishes. Some semi fluid like material will spread on the upper surface of the body. Some wounds will also appears in the body. Based on utilizing this chance, some bacteria and parasites will affect the fishes. To prevent the outbreak of any disease in the aquaculture system the pond environment should be kept neat and clean. Feeding should be optimum and split into 4 or 5 times and given in required quantity. Adequate oxygen level should be maintained in the pond by providing aeration.

7. Treatment Methods

If it is a nutritional deficiency disease, then that disease can be identified by which nutrient is deficient and that specific nutrient quantity can be increased in feed to overcome the disease. If it is other than nutrient deficiency disease then one of the following methods can be used to treat the diseases.

i)   **Bath treatment:** This treatment can be used to treat the disease which was present on the body surface of the fishes. Generally dangerous germs will be present in more numbers on the skin and gill surfaces of the fishes. Hence, the particular antibiotics can be mixed in water and the affected fishes should be allowed to swim in that antibiotics added water for sometime. This bath treatment can be given for hatchlings and also for brooders. To treat the fishes, it should be stopped feeding before 24 hrs.

ii) **Treatment through feed:** If the disease causing organisms were present inside the fish body, then this treatment can be done. It is done by adding particular quantity of drugs in the feed. When the fish take this feed, it can be treated through these antibiotics. This drug will destroy the bacteria which is present inside the fish body. This type of treatment through feed has some difficulties. Because, generally the disease affected fishes will not take the feed properly and they require immediate treatment. So initially if any disease affected fishes were found, then it should be observed in the lab to identify which type of micro organisms affected these fishes, and according to the identified microorganisms treatment should be done. If it is not able to identified, then oxytetracycline antibiotics can be used to treat the diseases. When the pellet feed is used, then drugs should be initially mixed with vegetable oil and then it should be mixed with feed.

If the treated fish is needed to sell, then after 3-4 weeks of treatment only it should be marketed. Before that it should not be sold for commercial purpose.

iii) **Treatment through injection:** For treating the fish brooders, this method can be followed. This method cannot be followed in cultured ponds. This method can be done by separating the disease affected fishes and injection can be done below the base of dorsal fin. Handling of fishes during injection is very important. Since catfishes have spines, it will cause pain those who handling the fish for injection. So larger fishes should be given anesthesia before injection.

Thus, this treatment is very useful for the fishes at initial growing stage. It is advisable to avoid the treatment very often. The commonly used drugs are given in table.

Table 14.1. Different Compounds and treatment regimes

Sl. No.	Compound	Treatment method and dose rate	Disease or Agents treated
1.	Formalin	Bath : 20-45 min, 100-250 ppm	External protozoa and monogenetic tremetodes
2.	Common salt	Bath: Indefinite, 0.1-0.2% Bath: 20-30 min, 3.0%	*Saprolegnia*, external protozoa, Crustaceans, leeches
3.	Trichlorphon (Dipterex)	Bath: Permanent, 0.25 ppm	Crustacea, Leeches, Gyrodactylus
4.	Malachite Green	Bath: Permanent, 0.1 ppm,24 hr	*Saprolegnia*, external protozoa
5.	Copper Sulphate	Bath: Permanent, 0.2-2 ppm. Do not use in soft water	External protozoa
6.	Potassium Permanganate	Bath: Permanent, 2 ppm Repeat treatment may be necessary	External protozoa, monogenetic trematodes
7.	Nifurpirinol (Furainace)	Bath: 1 hr, 1 ppm Bath: Indefinite, 0.1 ppm	Myxobacterial infections
8.	Magnesium Sulphate	Food: 3% of ration	Intestinal helminths
9.	Nitrofurans	Food: 10g/100kg fish/day for 10 days	Systemic bacterial infections
10.	Oxytetracycline	Food: 7g/100 kg fish/day for 10 days Injection : 50mg/kg	- do -
11.	Sulphonamides	Food: 15g/100kg fish/day for 10 days	- do -
12.	Potentiated Sulphonamides	Food: 5g/100 kg fish/day for 10 days	- do -

The following environmental and stress management practices can be done to reduce the incidence of diseases in aquaculture.

- Maintain the highest possible water quality

- Maintain prudent stocking densities and standing crops

- Disinfect the water supply and equipment used in the culture facility
- Expedite removal of dead and moribund fish
- Handle the fishes gently during stocking, sampling, etc
- Use prophylaxis during and after handling to aid wound healing

Diseases are also better managed by understanding the requirements of the disease agent. Fish hatcheries use saltwater at 5-10% to control ciliate protozoan parasites. Flatworm may be controlled by the use of low salinity water.

REFERENCES

A.K. Singh and W.S. Lakra. 2012. Culture of *Pangasianodon-hypophthalmus* into India: Impacts and Present Scenario. *Pakistan Journal of Biological Sciences, 15: 19-26.*

Bhujel, R.C. 2012. A Manual for Tilapia Hatchery and Grow-out Farmers. Aquaculture and Aquatic Resources Management, Asian Institute of Technology, Bangkok, Thailand. 67p.

Hussain, M,G. 2004. Farming of Tilapia: Breeding plans, Mass seed production and aquaculture techniques.149p.

Karal Marx, K. and N. Ramanathan. 2004. Catfish Culture (Training Manual). NATP. TANUVAS. 66p.

Karal Marx, K.2007.Technology Development for the production of Monosex and Triploid catfish *Clariasbatrachus*using chromosome manipulation techniques.ICAR Scheme Final report. 54p.

Karal Marx. K, Sugumar, G., Kiruthigalakhsmi. D and M. MuthuAbishzag. 2016. Quality seed production Technologies of Carps, Catfish and Tilapia. 104p.

Karal Marx. K. 2017. Manual on Tilapia Farming. Tamil Nadu Dr. J. Jayalalithaa University. 128p.

Nandi, S., N.K. Barik, P.R., Sahoo, P.N, Ananth, Ambika Prasad Nayak, J.K. Sundaray and P. Jayasankar. 2014. CIFABROOD ^{TM} as Carp Broodstock Diet: Experience from farmer's pond. ICAR-CIFA, Odisha, Brochure.